Ecology, Spirituality, & Education

Studies in the
Postmodern Theory of Education

Joe L. Kincheloe and Shirley R. Steinberg
General Editors

Vol. 201

PETER LANG
New York • Washington, D.C./Baltimore • Bern
Frankfurt am Main • Berlin • Brussels • Vienna • Oxford

Elaine Riley-Taylor

Ecology, Spirituality, & Education

Curriculum for Relational Knowing

PETER LANG
New York • Washington, D.C./Baltimore • Bern
Frankfurt am Main • Berlin • Brussels • Vienna • Oxford

Library of Congress Cataloging-in-Publication Data

Riley-Taylor, Elaine.
Ecology, spirituality, and education:
curriculum for relational knowing / Elaine Riley-Taylor.
p. cm. (Counterpoints; v. 201)
Includes bibliographical references and index.
1. Human ecology—Social aspects. 2. Knowledge, Theory of.
3. Education—Philosophy. 4. Ecofeminism. 5. Postmodernism.
I. Title. II. Counterpoints (New York, N. Y.); v. 201.
HM856 .R55   304.2—dc21   2002004100
ISBN 0-8204-5543-1
ISSN 1058-1634

Die Deutsche Bibliothek-CIP-Einheitsaufnahme

Riley-Taylor, Elaine:
Ecology, spirituality, and education:
curriculum for relational knowing / Elaine Riley-Taylor.
–New York; Washington, D.C./Baltimore; Bern;
Frankfurt am Main; Berlin; Brussels; Vienna; Oxford: Lang.
(Counterpoints; Vol. 201)
ISBN 0-8204-5543-1

Back cover photo by Jim Taylor
Cover design by Lisa Dillon

The paper in this book meets the guidelines for permanence and durability
of the Committee on Production Guidelines for Book Longevity
of the Council of Library Resources.

∞

Printed in the United States of America

Guide me in this writing—make it yours.
Help me know what direction to take, so that this work
will have your "life" through it.
Do you want me to speak for the earth?
For poison anthills and mallard lives crushed on roadside concrete?
For smoke-hung sunsets and rattling, creaking
axles over interstate bridges?
Make your stand here if you want to, I'll write it with you.

Contents

Acknowledgments

The dance of the wind is one always in-negotiation. There is no a priori, no preconception of form or destination. The wind is perpetual movement through and across and between, affecting and affected in the process of its passing.

—Journal entry, Bergamo, 2000

A long continuum of living has led to a culmination of this project, although some phases of my own history come to mind as having been more strongly influential: turbulent early years that taught me there are "other ways of knowing" and being that are beyond those prescribed by status quo conventions; four years moving up and down the western coast of the United States and wintering in desert ghost towns of southern California; and the last several years spent living and writing at the small cabin on Bayou Fountain in Baton Rouge. Spirit has been a constant companion throughout, drawing me on and connecting me with people who have lent to life a richness and who have helped create a context enabling my life's passion to be brought into the arena of my life's "work."

In this regard, I begin by thanking those who have been vital in the fulfillment of this project: Wendy Kohli, an exemplary scholar who has deftly walked the fine line between being both mentor and friend. I can't express what her caring guidance and steady reserve of strength has meant in seeing this work through to "completion"; William Pinar, whose vital presence, coupled with a quiet dignity, has encouraged me to look within myself and create a "running of the course" that is uniquely my own; William Doll, whose creative fusion of postmodern process theory with curriculum vision has been a source of inspiration since the first day we met; and Petra Munro, who has supported me with her time and her insights, always challenging me to think beyond the well-worn path toward horizons new and foreign at first glance—but which have become vital to my understanding of what it means to "do" curriculum; never did I imagine curriculum theory could be so resonant with *who I am*. For having come into relation with this unique and passionate body of scholars, I am most grateful.

I extend thanks to many who have been close by through the writing process: heartfelt thanks at the outset go to my parents, H. K. and Laura Kent, for all that they are and have meant in my life, each amazing individuals of whom I am most proud. I thank my brother and his wife—Gerry and Vicki Kent—ever supportive and instrumental in initiating the desire to begin this education journey. I owe a special debt of gratitude to my sister Barrie, who has encouraged and supported me emotionally, intellectually, and physically all along the way, and also to her husband Fern, whose kind and generous spirit have been forthcoming on every level. As well, I thank Richard Keeler, who demonstrated a faith in me both intellectually and spiritually, and who took it upon himself to sponsor my education through the beginning years of this process. To all of you, I express my gratitude, certainly for the part you have played in making this possible, but more than that—just for being part of the fabric of who I am.

Many thanks to Karolien Debusschere, who stood by me and guided me through many phases of this experience. Most sincere appreciation to Vikki Hillis and Larry Erikson, to Hampton Peele, Charles Flanagan, Kane Mire, Debbie Maglone, Dona Parker, Heidi Beall, Jack Westbrook, Debbie Allen, Jacob Wolfe, Kelvin Broad, Carolyne White, Bea and Don Jacobs—each of you are unique particles of light on this planet, and it is an honor to call you my friends. As well, I extend gratitude to my "new" family, Tori, Mel, Jordan, and Amelia Percy, and also to Trey Taylor, who has taught me the meaning of unconditional love. Thanks also to Rambo, who has been close at hand through this last year of writing.

Special gratitude to those friends and comrades who have walked with me through the graduate school process: Molly Quinn, Marla Morris, Jennifer Falls, Peggy McConnell, Al Alcazar, Steve Triche, John St. Julien, Kimberly Vannest Callicot, Denise Taliafero, Toby Daspit, Lysah Kemper, Doug McKnight, Tayari Kwa-Salaam, and Jeff Gagne—the world is very small and though we go in myriad directions, I know that we remain connected by a special bond of what we have come through. Also to Louis and Beth Reames for making respite at the cabin possible for myself and so many before me.

Among the people I give credit for seeding my penchant for connecting spirit into everyday life are those kindred beach-dwellers at "Seaside" north of San Diego: Dana Eddy, Bob Turgeon, Raven, Annie and Greg Hogan, and others who shared meals and long days-into-nights before the roll of surf and the ritual sunsets during those year-round summers near Cardiff-by-the-Sea.

Thanks also to Naked Lunchers Sharon and Brian Andrews, Scott Chauvin, Russ Seal, Carlo Cuneo, and Tommy Dupuis. You brought music back into my life and provided an outlet for creative release at times when my reality would become a bit too cerebral.

I wish to thank Donna and Larry Pierce, and also Frank Parker. Each encouraged me to turn a corner in the direction of my heart's passion with regard to my theoretical focus. I never imagined the richness that would ensue from such a shift. I am grateful to have had many spiritual mentors—some I have already mentioned—but I take this opportunity here to acknowledge Russell Riley, who over many years of sharing my path introduced me to myself. As well, I wish to thank Mark Courtman, Melinda Walsh, Marge Kass, and Jan Mahannah-Moskus for always being there and for consistently reminding me to return to the prayer of my heart. Finally, gratitude beyond words to my closest friend and partner Jim Taylor for his love, encouragement, and unwavering resilience through the peaks and valleys of this writing/living experience. Each day I stand in awe at the magic of life's process for having woven our lives together into a richly layered synthesis. Thank you for being in my life.

Special thanks to Shirley Steinberg and Joe Kincheloe, and to those at Peter Lang, who invested time and energy to bring this book to life.

Permission to reprint the following excerpts is gratefully acknowledged: Riley, E., "Educational Awakenings on Relational Ways of Knowing." In Oldenski, T. and Carlson, D. (Eds.), *Educational Yearning: The Journey of the Sprit and Democratic Education*. New York: Peter Lang, 2001, pp. 136–159. All rights reserved.

Also, Riley, E., "Deep Ecology: Reflections on Holistic Education as Praxis." Baton Rouge: *National Journal of Wellness*, 1998. Reprinted by permission of Golden Tradewell, publisher, pp. 27–32. All rights reserved.

Introduction

No Thing in Isolation: Awakening into Process

Watching from my cabin window the stark, cold January morning stands in silent relief against refrains from earlier seasons. Humming sounds of summer insects skirt along the bayou separating a grassy apron below from the lush Louisiana woodland beyond. Green summer tangles of vines and leaves are cast into colors of gold-to-orange-to-brown with autumn's rustling, cooling winds. Angles of light shift with the seasons. Shadows move and dance in a play of particularities, relational nexus[1] of light-to-object in time and space. Life is cyclic, it is staccato, it is moving and ceasing to move, birthing-living-dying-birthing-ever-again. Commingling processes shape and are shaped by their interconnections, evoking a complexity of forms rich beyond belief.

But the wondrous quality of this richness can be invisible, also, within its own continuity, within the ways that it is ever with us, ever-changing but ever-present.[2] We often cease to see it, to hear it, to feel it. We grow numb to it. I know that this is true for me. I often fail to feel my own connection within that complexity, as if my sensibilities were clouded over and I were separate from the immediacy of experience. Herein lies my personal longing for an engagement with the "spirit" of life, for an inner recognition of my own embeddedness within the dynamic process of living. At times I have glimpses, experiences of "knowing" my place within that process—I seldom look out on the bayou behind my house without a sudden awareness, a breathing in, a connection felt with my surroundings, the larger biosphere of which I am a part.

I want to speak of this connection, because I feel it must be spoken—for our children, for our people, for our planet and its peace. It is a call worth hearing, its message mirrored on the pages of our culture. As humans, we must find our way over landscapes mired with anguish, assaults, and assignations, inspired to keep moving toward some unnameable pull, a draw toward some perceived fulfillment, some final quenching of thirst. But, from where do we draw our water? How do we answer the question, What will bring peace and contentment? Is it money? Is it health? Is it power and control? Is it relationship, or beauty, or status and image? We race ahead to fill the longing, but does the water we drink quench the desert of our inner thirst?

Here is a call that yearns for what has been separated and one that also rings with a joy at the resounding rhythm of life. Possibilities are all around us—the empty, the full, and all that lies between—the choices among them are ours, a choice to wake up, a choice to hear the call of our own heartbeat. Surely we recognize the rhythm, the spirit that animates our deepest places and pulsates through every now moment, that rhythm which vitalizes our being from its prison of numbness. But do we feel that rhythm? Do we hear that call? As often as not a numbness overtakes us.[3] We live in a mechanistic[4] world that has taught us to think in terms of separation: inner self from outer self; self from others; self from nature and the planet. We have replaced inner rhythms with routines in many ways. We wake to an "alarm," live our lives according to the hands on a clock, spend most of every week at a job that, in many cases, is disconnected from an internal desire to create or to engage in a task for its own sake. So often, the meaning within our work is disconnected from the immediacy of experience, or is even unknown to us. Intrinsic motivation is replaced by extrinsic motivators such as output requirements and job security. The capitalistic system of the United States—coupled with industrial and technological growth—keeps us running on a wheel of ever-escalating production and consumption. Eric Fromm (1986) reminds us that, in our society, a person's self worth is more often based on *having* than on *being* (p. 20). What has come to drive the day-to-day routine for many in the United States is a desire for immediate gratification of material wants so as to increase prestige or personal standing on a scale weighed hierarchically against others.

Advances in technology have, likewise, increased our capacities to exploit and destroy the natural world. We pollute the rivers, the air, and the soil with chemicals, pesticides, and toxic waste. We level forestlands to provide paper products for our "disposable" society. Ecofeminist author Starhawk (1994) tells us that as humans we see ourselves "as the 'Crown of Creation' for whom the rest of nature exists"; meanwhile our resources are wastefully plundered (as cited in Weaver, 1994, p. 250). In her essay on the "consciousness of estrangement," Starhawk (1994) suggests that as Westerners "we do not see ourselves as part of the world—we are strangers to nature, to other human beings, to parts of ourselves" (p. 176). The "relational knowing" of which I speak is an ecospiritual[5] awareness framed by the relationship I share within the cosmological[6] world. It attempts to overcome the dualistic separations underlying Western cultural thought: separation of mind from body, individual from community, human beings from the natural world and from the larger planetary context out of which we arise as a species.

The sense of estrangement and alienation characterizes so many aspects of our culture, indeed of our lives. Perhaps we numb our consciousness in defense against the context in which we live. As if in order not to feel the pain of separation, we cloud over the sense of lived experience itself, the epitome of "be-ing" which Mary Daly (1978) has described as the verb that depicts "the dimensions of depth in all verbs, such as intuiting, reasoning, loving, imaging, making, acting, as well as the couraging, hoping, and playing that are always there when one is really living" (pp. 23–24). With Susan Griffin (1990) I feel that

> whether we want to or not, we share a social and biological matrix. We are connected. When we violate others or simply fail to feel this connection, we feel instead an emptiness, a mourning, an undefined grief. (p. 95)

Drawing on contemporary anthropology, Juanita Weaver (1994) notes that it is the function of every culture to establish the "individual's relationship to the self, the individual's relationship to others, and the individual's relationship to the universe" (p. 250). With this in mind, we might locate the numbing pain of separation within the inner longing to experience more fully the larger complexity of connections in which we are embedded.

As an educator, it is this sense of estrangement, of alienation, that I lament in our schooling. My project, then, is to explore a relationally based, ecospiritual perspective[7] and its implications for curriculum theory. If there is a place where we might seek to foster a sense of relationality—between the illusory divisions within our own being, with all that is around us in the social community, and in the natural world—it is here with the children. The future of our planet is in their hands.

An Awakening Praxis: Intentions and Commitments

Education is a "calling" for me. While that has only become clear of late, I recognize now that it has been "true" all along. Something inside, some unknowable knowing, has led me here onto this ground so familiar and yet so strange. The "curriculum field, like other academic disciplines, is a conversation" (Pinar et al., 1995, p. 849). My intention here is to step into that dialogue. My sense is that what I have to say can be said within the discourse of a "reconceptualized" field (Pinar et al., 1995), one which has moved beyond its former "institutionalized aim...to one with a critical, hermeneutical goal of *understanding* practice and experience" (Kohli, 1984b, p. xvi; see also Kohli, 1991c; emphasis added). As a teacher, the practice and experience I wish to understand is my own and in that way, perhaps, I may better assist other teachers in their own personal praxis, their own understandings (Smith, 1996). The search for me involves an exploration of my own "becoming," a process that is never finished, a continual path of creation toward that "moreness" (Huebner, 1995, p. 344), that "as-yet" (Greene, 1996), that ever-opening onto possibilities for an engagement with the spirit of living. For me, the focus of education—learning—is a spiritual journey. It is *creation* itself.

Education begins with each individual being, not only "turning inward" but also "moving outward;" a search for self and also self's relation within the larger frame of community, society, world. It is also about children and about those who would guide them toward a "becoming" of their own—into the fullness of life, the richness of relationships, the strengthening and broadening of skills and abilities. It is about nourishing their capacities for negotiation, discernment, and fairness, so that they may come to recognize their own responsibilities as members within a larger matrix of life—responsibility *of* being (Doll, 2002), of their own self-growth, their relationships, their communities, the planet. It is living and acting with an intention to create a better world, recognizing the responsibility it takes to be an integral part of something larger than ourselves. Critical, then, is educators' task to help children understand their *interdependence* within the ecological habitat from which humans draw their life and sustenance. My intent and my commitment is to foster this understanding, in myself and in others, through the educating process.

According to C. A. Bowers (1991), however, Western conceptions of individualism and freedom have placed emphasis on schooling as a medium for the "emancipation of talents, interests, ways of becoming" which have led to a "cultural orientation...exploit[ing] the habitat in a short-sighted and self-indulgent manner" (p. 327). He warns that

> the changes in atmospheric chemistry, increased acidification of lakes and soils, deforestation, extinction of species, contamination of marine habitats, and the build-up of solid waste...[will be] further exacerbated by a world population that has doubled from 2.5 to 5.0 billion in the last thirty-six years and threatens to double again within an even shorter time frame. (p. 327)

Therefore, an infusion of understandings grounded in ecological relations should be prominent within a child's education, beginning with an embrace of the idea of "holism,"[8] a recognition of the interconnection of all life, which is by my definition a *spiritual* awareness. Whether referring to the biological or the human community in which we reside, a personal search for the sacred within everyday living fosters a recognition of the interrelationality of all that exists. In addition to the cultivation of a reverence for all life, my hope for education is that those aspects of human knowing that have been marginalized as less important than the 'rational' center be recognized for the contributions they make to enriching and strengthening human understanding. This interrelational way of knowing is the foundation of my ecospiritual perspective (Berry, 1988; Fox, 1995; Kaza, 1993; Spretnak, 1997; Zukav, 1989). Ecospiritual knowing is a way of viewing the world relationally, an epistemology grounded in an awareness of truth as contextual and the individual as an integrative being.

We all have our commitments, our own agendas. When entering school buildings, we can only *pretend* to be neutral and leave our values at the door (Pinar et al., 1995). We have learned, alas, that pretense does not ensure performance or result. In fact, "performance" is perhaps a fitting term for so many decades of "bland" and lifeless lectures, pretending that the fragmented subject matters of schooling emanate in a smooth linear stream, as if from some sequestered source on high. Teachings "transmitted" as "truth" are legitimated by rationality as the knowledge worthy of passing on to the next generation.

My commitment issues from elsewhere. It is one wherein educational legitimation is extended to include not only "rationality" (i.e., that which is verifiable through the intellectual mind) but also that which can be learned through emotions, imaginations, intuitions, perceptions of wider vision, or understandings that cannot be pinned down and labeled or separated into categories of "truth" or "falsehood." Reason is but a *part* of humans' wider abilities to perceive and to understand ourselves in relation to others and to the planet. Our traditional educational model has long relied on a reductionist way of knowing, emphasizing the legitimacy of a *portion* of the full range of human capacities to know. Schools are places, quite often, where many come to feel a sense of alienation, of numbness, of separation from parts of themselves, from the larger community, and from the ecological world. As a result, children in mainstream schools have been de-educated, in many ways, taught to close their eyes to the wide array of sensibilities which vitally affect who they are and how they know. My commitment is toward "awakenings"—personal and global—both inside and outside of classrooms. By awakening, I mean to an awareness of the multitudes of ways that humans can be seen as relational beings, beings existing within an interconnected web that is more fully experienced through the integration of our perceiving senses, metaphorically, an integrative working of *body-mind-spirit* into a confluence of forces, acting, intersecting, and infused with a vital "life force." By awakening I mean an ever-deepening recognition of the sacred character of all life, sustained through a creative spirit that animates all living things and joins all upon the earth as an integrative body. And, by awakening I also mean a broadening cognizance by humans of the wonder of being alive as a part of that living, creative force, and of themselves as creators with the potential for agency.

Beginnings

The Buddha met a stranger on the road, who asked him who he was. "Are you a god?" "No," was the reply. "Are you an enlightened being?" "No," the Buddha told him. "Are you a magician, or a man who flies with the eagles? Tell me sir, please, what is it that you are?" The Buddha answered, "I am *awake*."

I admit there are times it seems humanity[9] is sleeping. By sleep I mean "a state of apathy or indifference...inactive...dormant...numb" (*American Heritage Dictionary*, 1996, p. 1695). Viewing the world through an anthropocentric lens, we humans have become apathetic to opportunities before us that could contribute to the healthful longevity of the planet that is our home. There is great potential for helping to affect a symbiotic balance between our species and the larger ecological world. However, most people remain largely inactive, with eyes closed to the potential role we could play as stewards of a planet inhabited by multitudes of life forms in an ecological balance. The word *steward* drawing from Old English means "watchful," "awake," from the Greek, "revere," and from Latin it means "to respect" or "to feel awe for" (*American Heritage Dictionary*, 1996, p. 1764). Unfortunately, those humans demonstrating respect or reverence for the earth stand in the minority compared with the many people who ignore, apathetically, the fact that we are part of a social and biological matrix.[10]

The human-earth relation exists within a delicate balance, yet we must learn ways to tread lightly, for we humans overstep our bounds. In his "land ethic," twentieth-century nature writer Aldo Leopold (1949) has suggested that we see the earth as a community, rather than a commodity. He defines an "ethic" in ecological terms by saying that it "is a limitation on freedom of action in the struggle for existence" and in philosophical terms, it is the "differentiation of social from anti-social conduct" (p. 202). He further says that

> [a]ll ethics so far evolved rest upon a single premise: that the individual is a member of a community of interdependent parts. His instincts prompt him to compete for his place in that community, but his ethics prompt him also to co-operate [sic].... The land ethic simply enlarges the boundaries of the community to include soils, waters, plants, and animals, or collectively: the land. (p.203– 204)

Implicit within this idea is a sense of human responsibility based on a deep respect for the earth and a recognition of our own embeddedness within the larger planetary context. If the planet doesn't survive, neither do we; however, as humans we tend to distance ourselves from the very medium from which we arise. We fail to see that we are poisoning our world, thereby poisoning our bodies, our people, and the environment that supports and sustains life on earth. These violations of planetary health are also violations of our own well-being, and could ultimately mean the end of humankind (Berry, 1988).

Alienation also pervades the field of education, fed by the rationalization processes (Weber, 1968; see chapter 1) institutionalized into schools, whether through the mechanistic, factory-model schooling of the early twentieth century or the more recent "corporatist reorganization" of schools (Wexler, 1996, p. 20). In both cases, possibilities for relationality and community among those who inhabit school classrooms is overshadowed by a focus on efficiency and control in the former, and on "performance-based outcomes," in the latter "a productionist emphasis," wherein "restructuring" and "reform" exists as "part of a wider process of social structural rationalization, instrumentalization, and corporatism"[11] (p. 19).

Educational Ways of Knowing

Curriculum scholars over this century (e.g., Dewey, 1956; Huebner, 1995; Macdonald, 1964/1995; Pinar et al., 1995; Zirbes, 1934) have contested an overreliance on rationalistic models within education, especially as they have manifested as industrial models of schooling (Tyler, 1949) or the mechanistic theories that frame behaviorism (Thorndike, 1913; as cited in Pinar et al., 1995). As is the case with most academic disciplines, the field of education has appropriated scientific method and objective analysis as foundational to our knowing. What has this reliance on rationality meant for the state of American education and for the children served by systems of schooling?

Behaviorism's Influence

A means-end frame for educating children was put forth early in the century by social efficiency experts seeking to apply the techniques of industry and behaviorist psychology to schooling. Nationwide interest in industrial growth and technology was reflected within the profession so that there was an increasing call for "efficiency in education" (Sequel, 1966, p. 67). Part of this emphasis was reflected in the creation of "methods for measuring aspects of education, often called the measurement movement," coupled with "the new psychological theorizing of Edward L. Thorndike" (p. 67). Thorndike sought to make education objective and verifiable and thus "adopt[ed] the research methods of the physical sciences" for the field of curriculum with a strong emphasis on stimulus-response behavioral psychology (Pinar et al., 1995, p. 91). Through the reduction of

> each human action to its smallest unit, that of stimulus and response, Thorndike...sought to establish the principles of human behavior that would permit its prediction. (p. 92)

Thorndike held that a child's behavioral response to a stimulus "indicated the content of a child's learning...[so that] response = learning" (p. 92). Such a quantifiable conception of learning allowed human experience to be studied scientifically and then mathematized so that "responses could be [statistically] judged for probability, compared, tabulated, ordered and correlated...for determining the effectiveness of teaching and learning" (Sequel, 1966; as cited in Pinar et al., 1995, p. 92). Meanwhile, the advance of "statistical research and measurement functioned to legitimate another emerging reform movement, the social efficiency movement" (p. 93).

Experimental meets Efficiency

In addition to the influences of experimental psychology and statistical analysis on the field early in the centuries, the idea of creating a more "scientific" view of educational practice was an outgrowth of the industrialization of U.S. cities. The late nineteenth and early twentieth century saw rapid immigration into fledgling U.S. cities from abroad as well as rural Americans flocking to urban areas in search of jobs. Lacking infrastructure to support these masses of people, "leaders, both lay and professional...[searched for] attractive organizational schemes to borrow from" (Tyack, 1974, pp. 29–30) and made rapid movements toward bureaucratization "to replace confused and erratic means of control with careful allocation of powers and functions within hierarchical organization" (p. 28). In an attempt to handle the socioeconomic problems which come with rapid growth and change, William Doll (1993) says,

> America turned to its schools and the model it used was that which made its factories productive—scientific management. Curriculum became a...national obsession; and the scientific curriculum was based on efficiency and standardization. (p. 48)

Educators garnered the principles of management for efficiency "found in the factory, the army, the newly created police department, and even the railroad" (pp. 29–30). Franklin Bobbitt applied Frederick Taylor's principles of scientific management to the business of schooling through the formulaic ways it was being used in the factory. Managers would identify and analyze tasks within a particular division of labor, and its component increments would then be "sequenced as work instructions" much as school subjects are sequenced and divided today (Pinar et al., 1995, p. 95). Bobbitt's influence did much to further situate the developmental "how-to" mindset for the systematization of factory model schooling where "punctuality, order, regularity, and industry [were seen] as essential features of a uniform urban discipline required for success later in life" (Tyack, 1974, p. 42).

Ralph Tyler deepened the hold of the widely accepted social efficiency vision of schooling through a "linear, administrative procedure for curriculum development" (Pinar et al., 1995, p. 148) which came to be known as the Tyler Rationale. The four steps of "1) chosen purposes, 2) provided experiences, 3) effective organization, [and] 4) evaluation...[were] but a variation of Descartes' general method[12] for 'rightly conducting reason and seeking truth in the sciences" (Doll, 1993, p. 31). The "techno-scientific" model of educating for order had been established so that within schooling there came to be an over-reliance on developing the rational mind to the neglectful omission of other important aspects of the human character (p. 54).

An instrumental means-end frame for educating children has influenced school structure and practice for several decades since the 1940s. With the launching of the satellite *Sputnik* in 1957, American education escalated its emphasis on science and math in the curriculum (Pinar et al., 1995, p. 154) as a move toward ultimately strengthening national defense. Knowledge was seen as being easily separated into component parts, fragmented into content areas, with a distancing separation of teacher from student in vertical relation hierarchically for the maintenance of control. This "observer consciousness" understood curriculum along very narrow lines (Kesson, 1994) with a view toward maximizing efficiency in learning and objectivity in evaluation. It was a way of educating that emphasized prediction and control.

The testing machine, so deeply entrenched, has driven the way that curriculum has been conceived (measurability) and the way that it has been taught (transmitted), in order to obtain objective results from which to verify that a predetermined body of knowledge has been "mastered." Schooling has emphasized the accumulation of testable facts and units of information, thus favoring the rational mind to the omission of other equally important, but less easily testable, aspects of a human being. A "transmission model" of learning has held that knowledge is "transferred from teacher to learner," and is a model "obsess[ed] with…measure[ment]" (Kesson, 1994, p. 4). The teacher has held a vertical relation above the student, as purveyor of that which is of "of value" (p. 4). As an authority who has "mastered" this "important" knowledge, the teacher could then transmit it to the passive recipient below—the student—in a one-way transaction with learning driven by an external locus of control. Observer consciousness could also be described as a "spectator view," where knowledge has been seen as being outside and separate-from the student, instead of being *constructed with* the student (Doll, 1993, pp. 168–169) within a relational context of mutual dialogue.

Analogous to the teacher-as-transmitter frame is Paulo Freire's (1970/1993) critique of the "banking system" of pedagogy, in which education is "an act of depositing, in which the students are the depositories and the teacher…[the] depositor" (p. 53). In this dichotomous scenario, teacher stands in opposition to the student as possessor of knowledge justifying his or her existence by the student's "absolute ignorance" (p. 53). This "false assumption" results in a distancing of teacher from student, human being from human being. It renders the student as object and contributes to what Fromm (1986) describes as "necrophilia"—a "propensity for death"—which increasingly characterizes our culture, as contrasted with the term "biophilia," a "propensity for life" (pp. 112–114). Fromm describes the

> necrophilous person" as one who "loves all that does not grow, all that is mechanical…He [sic] loves control, and in the act of controlling he kills life" (p. 58; as cited in Freire, 1970/1993)

Embedded in the controlling structure of Freire's banking concept is the notion that good students think and act as they are told, excluding the kinds of practices that encourage within students a critical "consciousness-raising" (*conscientizacao*) and that seek to empower students toward transforming their world (pp. 54–56). According to Freire (1970/1993), "action and reflection upon the world in order to change it" (p. 14), which he terms "praxis," may be fostered through pedagogy. This idea has been picked up by bell hooks (1994), who couples the idea of praxis with Buddhist monk "Thich Nhat Hanh's philosophy of engaged Buddhism" (p. 14). The latter, she says, would see students "as whole human beings with complex lives and experiences rather than simply as seekers after compartmentalized bits of knowledge" (p. 15). In this 'participatory' frame of educational practice, students and teachers co-construct knowledge in an atmosphere that honors both as being vital parts within a mutually constitutive relationship.

Educational scholars have long questioned "scientistic" views of schooling (see Dewey, 1956; Doll, 1993; Huebner, 1995; Macdonald, 1964/1995; Pinar, 1975). Many have encouraged movement away from "isolated, atomized 'observer' consciousness into relational thinking and being, or 'participating' consciousness" (Kesson, 1994, p. 2) and away from the mechanistic frame of fragmentary practices which arose out of the social efficiency and behaviorist models of schooling prevailing for much of this century.

Constructivism and Corporate Change

More recently schools are signaling shifts toward constructivist approaches to pedagogy that stress "students' active participation in the construction of knowledge and meaning" (Morrison, 1997/2000, p. 512). Constructivism works on the premise that student learning will be more effective when students construct knowledge from learning environments that draw on what is relevant to their particular interests (Goodman, 1986), styles of learning, and frames of intelligence (Gardner, 1993). Students are more actively involved in schools today with curriculums that emphasize interaction and participation in areas such as decision-making, problem-solving, self-discipline, cooperative learning, creative expression, and application of math skills to "real life," (Morrison, 1997/2000, p. 513). School and curriculum reform initiatives are widespread.

The factory model school has shifted somewhat in the postindustrial age, according to Philip Wexler (1996). He describes the most recent currents of reform in education as "corporatist reorganization," referring to the "progressive, liberal platform of educational reform and 'restructuring,'" which, he says, "represents a partnership of the state, business corporations, and significant groups of educational professionals" (p. 20). The trend is to replace the assembly-line model of educational practice with high-performance restructured workplaces that feature a "shift from an 'academic' to a 'real world' focus" (p. 26). While the broad changes characterizing constructivist reformed schools sound promising, Wexler maintains they are largely manifestations of a corporatist (economist) vision with a productionist emphasis, a momentum he interprets as yet another process of rationalization and instrumentalization within education (pp. 21–33). The new schools stress movement away from an assembly-line model, with emphases on flexibility, teamwork, collaboration, and inter- and intra-personal skills. However, a common thrust of the restructuring literature is a "*translation* of work skills into academic curriculum...[e.g.] productive use of resources...the curriculum as high-performance workplace...relevant skills taught in 'the context of real life situations and real problems'" (Department of Labor, Secretary's Commission on Achieving Necessary Skills; as cited in Wexler, 1996, p. 25; emphasis Wexler's). Metaphorically, schools have moved from the factory floor to the boardroom, as corporate "restructured workplaces," establishing "education as both more closely linked to economic production organization and as a social form directly analogous to new modes of production" (p. 22).

Autonomy and Androcentrism
Aside from a trend toward the capitalist commodification of education, Bowers (1995) says that underlying assumptions implicit in the new schools are based on liberal notions of the autonomous "individual as the epicenter of the universe" (p. 7). He sees the human-centered focus of constructivist education, which bases learning on the relevance of the child's own interests and experiences, as reinforcing the anthropocentric cultural beliefs and practices that have contributed to the ecological crisis that escalates around the globe. Bowers is clear that the "upward growth curve that characterizes consumer habits and forms of technological

development in modern cultures cannot be reconciled with the downward curve in the viability of natural systems" (p. 19). He says that what should measure the success or failure of a learning situation is its potential for socializing youth toward cultural beliefs and practices that will be ecologically sustainable over the years to come (pp. 5–6). Bowers is skeptical of constructivist theories of intelligence, such as Howard Gardner's (1993) theory of multiple intelligences.

In *Frames of Mind* (1993) Gardner delineates a broad range of human capacities for knowing, such as linguistic, logical mathematical, musical, spatial, bodily-kinesthetic, and personal knowing of self and of others. Understanding intelligence in this way, Gardner says, should enable psychologists and educators to identify intellectual strengths and "proclivities" (p. 10) early in a child's life, and then use those results to design educational programs suitable to his/her particular needs and abilities. An advantage of Gardner's theory of intelligence, Bowers maintains, is that it may "challenge...prevailing orthodoxy that represents intelligence as an attribute of the autonomous individual" (p. 98). However, in that it doesn't complexify the individual as being in relationship with the environing world, it "simply expand[s] the way educators understand the attributes of the individual" (p. 98), thus maintaining the status quo. Bowers prefers a theory of *"ecological intelligence"* which abandons "the Cartesian representation of the individual as spectator of an external world," for one wherein "the individual...[is] an interactive member of the larger and more complex...culture/environment relationship" (p. 15). In emphasizing the relational embeddedness of individuals, educators contribute to the development of a child's "sense of identity that incorporates the multiple relationships...that make up the environment we share with other members of the biotic community" (pp. 27–28).

Individualism and Ecological Sustainability

While I support constructivist theories that encourage the active engagement of the student in her/his own learning process, I agree with Bowers that the focus on learning as the zenith of autonomous individuality can too easily slip into omission of the importance of the relations—the family, school, community, and ecosystem—in which the child is embedded. In its coupling with capitalism, 'corporate' schooling can easily become another process of bureaucratized rationalization and instrumentalization wherein the bottom line is student achievement, based on suitability to perpetuate an economically restructured global marketplace. The liberal "individually-centered" (Bowers, 1995, p. 109) consumer lifestyle of the late-twentieth-century United States is inscribed within the guiding metaphors that shape privileged Americans' ways of knowing, valuing, and living life. In order to ensure the viability of natural systems, humans must learn to evaluate the beliefs and assumptions encoded in their cultural metaphors in order to shift from self-centered to eco-centered worldviews. For this reason, I believe that serious reevaluation of the ideology of individualism that underlies anthropocentric, androcentric, and Eurocentric worldviews in this country will be a necessary part of an ecological perspective on curriculum. An eco-centered view would necessarily bring with it a recognition of "the fundamental interdependence of all phenomena and the embeddedness of individuals and societies in the cyclical processes of nature" (Capra, 1975/1991, p. 326).

Educators are becoming more aware of our responsibility toward helping young people meet the challenges of a world which cannot sustain the continued mis-use of natural resources driven by a narrow, anthropocentric view. It is time for a curriculum approach that examines the "patterns of ecologically sustainable cultures and teach[es] students how to recognize and participate in the noncommodified activities of their own communities" (Bowers, 2001, p. 106). Developing skills and abilities toward becoming active producers in the workforce is only a small portion of what is important for educating children. Education should address a wide range of human potentials, fostering within children an awareness of their own human capacities for wisdom, imagination, appreciation, and indeed, their own responsibility within something which is larger than themselves: the family, the community, the ecological world (Noddings, 1992).

Therefore, my project explores the possibilities for an ecological vision of education based on relationality, an *ecospiritual praxis* fostering within children a broadened, deepened sense of their connection within the matrix of life constituting the earth. With Wexler (1996) I realize that education far exceeds the concept of schooling. I offer my vision of an ecospiritual praxis as a heuristic for walking a path in active engagement, self-reflection, and intention to sow seeds of life-giving change upon the planet. I agree with Kathleen Kesson (1994), who asks "[h]ow might we begin to think differently" about curriculum. Could we bring curriculum to the fulfillment of what she describes as a spiritual function by

> adapting to the unpredictability, the idiosyncracies, the dynamic process implied in such a model? Could we cope with the novelty that would be introduced into our systems? Might we begin to think of curriculum, as George Willis recently suggested, 'as an occasion for drawing the finite closer to the infinite?' (p. 5)

Kesson distinguishes between the spiritual and the religious by saying that there is a need for an invigoration and infusion of the human spirit into secular life and a reappropriation of the notion of the "spiritual" without the implications and institutionalized assumptions that surround the idea of "religion." She (1994) notes that unlike the term "religion," which tends to "stress the ultimacy of categories such as 'matter' and 'creator,'" the term *spirituality* is used here to "emphasize the actuality of process and self-creativity" (p. 3) within both curriculum and classrooms. Matthew Fox (1995), a theologian and former Dominican priest, attributes some of the baggage of Western religion to the dualistic tendencies of patriarchal worldviews. In this frame, spirituality is labeled "womanly"

> in the pejorative, dualistic, and patriarchal use of the term *womanly*...[as] passive and inert...a useful thing for a massively patriarchal society to accomplish...it allowed patriarchy to run wild with its militarism and war games,...bloated left-brain definitions of schooling,...rape of Mother Earth,...[and] its replacement of authentic worship—which is always a matter of relating microcosm to macrocosm—with words. (p. 7)

I agree with Thomas Berry (1988) that "what is needed...is the deeper meaning of the relationship between the human community and the earth process" (p. 10) so that we may cultivate "our sense of gratitude, our willingness to recognize the sacred character of habitat, our capacity for the awesome, for the numinous quality of every earthly reality" (p. 2). "Separation" as a way of knowing has kept us from experiencing how deeply our lives are interwoven within the fabric of the biological world. As an educator, I see strong possibilities toward a "reweaving" through a large-scale effort at consciousness-raising among people, particularly children in schools. I hold to a vision of education as a prime medium in which to initiate an educational praxis which draws on ecological and spiritual tenets of relationality and connection-making.

Re-envisioning education from an ecological perspective will foster within children an awareness of and respect for all living things as being sacred in the fulfillment of the creative process within which we all reside, and will encourage an ever-widening and deepening understanding of political and global relations (Blumenfel-Jones, in press, b; as cited in Pinar et al., 1995; Kohli, 1996). My ecospiritual perspective on education considers ways that the work of curriculum theorists links with theology, east and west (Fox, 1995; Ruether, 1996; Suzuki, 1970/1998; Wexler, 1996), particularly as that overlaps with deep ecology, ecofeminism, and process theory.[13] I am interested in work that critiques the anthropocentric worldviews denying the interconnections between human culture and the ecological world. Worldviews based in separation[14] are strongly implicated in initiating and escalating environmental crises.

Deep ecology is the term Arne Naess term coined in 1973 that describes the "deep long-range" vision to "examine, question and try to change the value systems and worldviews which are the ultimate causes of the external environmental crisis" (Bragg, 1999). Describing an ethos driving the deep ecology movement, Bill Devall (1988) says that in "grounding ourselves" within the "experience of our connection to the earth...supporters of deep ecology are fighting against thoughtless and mindless behavior" (pp. 11–12), and advocating eco-centric (or earth-centered) worldviews. While deep ecology has valuable insights to offer, I agree with global ecofeminist Ariel Salleh (1997), who suggests that deep ecology fails to critically examine androcentric components of anthropocentric worldviews and their "masculinist assumptions" (as cited in Merchant, 1996, p. 205).

Ecofeminist theory makes a vital contribution toward problematizing supposedly gender-neutral concepts of modernity that are based in androcentric separation exemplified by hierarchal, power-over (Eisler, 1990), patriarchal ways of viewing the world. In many instances, the inherent separation in power-over thinking denies the possibility that there could be a deep spiritual connection holding all things upon the earth within a network of mutually sustaining relationships. We can look to Buddhist/ecofeminist theorists (Kaza, 1993; Macy, 1991a and 1991b; Warren 1993;) in conceptualizing an ecospiritual vision supported by the notion of *systems thinking*, the idea that life's dynamic process consists of an interconnected web of relations. I am particularly drawn to ecofeminism, as it opens the door to other ways of communicating that match my own experience (Kaza, 1993, p. 54) and knowledge, situated within an accumulation of particularized experiences of what spirituality means to me tempered by my own gendered, raced, and classed background. As does the wider umbrella of feminism, in general (Kohli, 1993), ecofeminism allows me to bring forth my *own* voice, "speak [my] *own* truth" to reclaim the story of what I "know from direct experience" (Kaza, 1993, p. 5; emphasis added).

An ecospiritual view is also supported by many process scientists. Physicist Fritjof Capra (1975/1991) posits a view in which "the human spirit" is recognized as a "mode of consciousness in which the individual feels connected to the cosmos as a whole," thus is an "ecological awareness...[that] is spiritual in its deepest essence" (p. 326). He places this "perception of reality" as going

> beyond the scientific framework to an awareness of the oneness of all life, the interdependence of its multiple manifestations, and its cycles of change and transformation. (p. 326)

I believe that the symbiotic balance that exists within all of nature is sacred in character and worthy of our reverence. Writing on physics and religion,

Gary Zukav (1989) maintains that "reverence is...the experience of accepting that all life is, in and of itself, of value" (p. 51) Yet, through anthropocentric indifference many humans fail to show respect for the delicate balance of the earth as an interactive system of relationships within which we're a part.

I contend that we are connected, even though I sense a numbness, a separation dividing me from you. I trace this back to the ways that Westerners have come to privilege *thinking* over a more integrative knowing that combines mind with body, emotions, and spiritual under-standings, so that we might experience with more intensity the full capacity of our abilities for being and becoming in the world. It is a condition of our consciousness,[15] I would say, that limits our perception by narrowing it within a framework predefined by the "rational" model so valued in Western culture. A pervasive characteristic of modern "rationality" is a sense of alienation based in dualistic separation, a separation evidenced in our thinking. So often thought is directed and limited into predetermined binaries dividing outer from inner, me from you, male from female, mind from the body, or from the heart, or from the sacred. This alienation strips me of the fullness of my own *integrative* being.[16] There are times when I feel disconnected from all that is outside of me, from the community that surrounds me, and from the ecological world of my own genesis.

Deep ecology, ecofeminism, and postmodern process theories support the particular understandings of spirit that have been generated through my own experiences to shape my personal worldview. My own life's journey has brought me to an understanding which integrates all aspects of reality—every shade, form, and nuance—under a spiritual umbrella of interconnected relations. Spirituality is somewhat difficult to discuss without speaking in universal terms, since the meaning of the word *spirit*, for me, connotes the presence of a connecting universal principle uniting *all that is* in a constant of creation.

My vision of the Sacred is not a totalizing one, however. It recognizes the many within the one and also the one within the many (James, 1907/1995) as each is fundamental to the integrity of a larger order. I do not claim to understand what this larger order is, nor do I label it definitively.

To me it is a constantly emerging, generating, creative force, begetting life in all its forms, with its *own* life in a state of renewing genesis and continual evolution. I call it Spirit, Mother/Father God, "the Universe"—or I call it nothing at all, feeling hesitant to name it in any manner, for naming brings with it a danger. Labeling, or even speaking of a higher power, I feel, need be done with great care, because to pin it down with definition and determinacy can have an extinguishing effect. There is also need for caution and consideration so as not to diminish the ways others name, interpret, or conceive of its existence—or non-existence. I honor the right of each person to interpret this "force" in a personal way.

Spirit has been defined elsewhere, as "a causative, activating, or essential principle; the vital principle or animating force within living beings" (*American Heritage Dictionary*, 1996, p. 1737). From the Latin it means *breath,* which resonates with my own sense of the term as a vital force which "breathes" through all things, imbuing them with an animating force of life. I believe spirit is the force through which the human species is connected with the body of the earth, uniting all that exists within the vital breath of spirit's expression. The ecospiritual perspective to which I am committed, is "an ecological, earth-oriented, postpatriarchal spirituality" (Capra, 1975/1991, p. 340).

Implicit with this ecospiritual perspective is the practice of "mindfulness." Mindfulness is an awareness of one's positioning within the "now" as a site where action may be taken. The cultivation of mindfulness is a discipline subscribed to by many followers of Buddhism[17] in moving toward a state of being, which I see as akin to the sense of "awakening" I have described above. An ecospiritual perspective is one that acts with intention, with mindfulness, recognizing the sacred nature of day-to-day living and the potential for each human to create her/his world, working within the present moment. By dwelling on the past, or projecting into the future, humans lose sight of their "now," and yet it is the present moment within which decisions based on ethics such as Leopold's are possible. It is at the present moment that living really happens—the nexus of relations—the position where we may take actions, in and through the moment-by-moment experience of *living* life.

What is it here within the "now" that holds meaning and value for human beings? Granted, "value" may be different for each one, but, speaking for myself, I can always go back to the sense of immediacy I feel from my connection with the natural world, as if something pulls my gaze through this window glass and out to the tangle of brush just beyond. What is it that draws my sensibilities toward those upright figures of water oak and cypress with their sun-tinged trunks shimmering in the early-morning sun? What is it that moves my eyes toward the glistening needles of pine swaying and moving with the play of fat squirrels wrestling among the branches? What causes my sensibilities to liven with the tones and rhythms of life cascading like water over creekbeds and boulders. I attribute it to the deep recognition that I am a part of life's "process." The idea of the processional nature of life (Bateson, 1979; Capra, 1975/1991; Doll, 1993) ties in directly with my sense of spiritual connection as a vital part of the larger organic body of the earth, a part-to-whole relation, a micro-macro-cosmic view (Capra, 1996; Fox, 1988; Spretnak, 1997). Therefore, I would like to explore the possibilities of a relational view on curriculum that recognizes the processional nature of living as a basis for reconsidering how we might begin re-thinking, -feeling, -experiencing educational ways of knowing that are capable of supporting human individuals, their social communities, and their ecological environment with life-sustaining beliefs, values, and practices (Bowers, 1995).

Mapping the Contours

This work is divided into two parts following the introduction: the chapters in part 1 explore the ground of our knowing. Specifically, I analyze how problems emanating from an epistemological tendency toward separation have led to an ontological crisis of the human being that manifests, in part, as personal and cultural alienation affecting education. I question how the tendency toward separation has led to harmful anthropocentric and androcentric practices, conflating nature and women within patriarchal discourses and subordinating them into the status of "resource."

The introduction explores how "knowing" based in separation leads to many of the problems within ecology, education, and the human spirit. It discusses ways that separation is reinforced by modern "rational-ism," and the rationalization process, and how North American public schooling exemplifies just such a process (Weber, 1968; Wexler, 1996). Chapter 1 looks at the condition of modernity and how it is being questioned by postmodern and other anti-foundationalist theorists. I begin making a case for moving in between discourses of varying theoretical positions, e.g., those of postmodernism, ecofeminism, deep ecology, and process science to disrupt the "totalizing"[18] effects of using only one theoretical base. Chapter 2 continues the discussion of epistemology by delving into other ways of knowing, suggesting ways that schools tend to focus on the rational mind, while they have been known to discount knowing of the emotions, of the body, or of the spirit. The chapter sows seeds in beginning a discussion of an ecospiritual praxis as a basis for education. It also introduces the reader to the orientation of the author as justification for combining expository and autobiographical scholarship, a means of combining theory with personal voice toward a more integrative analysis.

The overall discussion in part 2 looks toward the work of "reweaving" modernist dichotomies, beginning with a discussion of difference. It then moves on to relational "ways of knowing," exploring what more holistic and ecologically informed perspectives would lend to curriculum in order to support sustainable life on a planetary scale. Through the lens of "difference," Chapter 3 draws on ecofeminist perspectives vital to a critique of both limitations and potentials of relationality as a heuristic for rethinking patriarchal "separatist" worldviews. Chapter 4 continues the discussion, drawing on the work of ecofeminists, deep ecologists, and process theorists to articulate a much-needed ecospiritual ethic and how that ethic can lead to an ecospiritual praxis. The work concludes with Chapter 5, which looks at ways to draw upon the idea of relationality in creating an ecospiritual praxis fueled by commitment, where factors such as intentionality and choice make agency possible. It begins by reconsidering relationships in which humans are embedded: those among our own varied interior regions as being integrally seamless and vitally connected, immersed within a sense "place," within communities that are multi-layered and ever-changing, and which provide cohesive mediums wherein all life forms partake in the warming fire of the "social." The chapter envisions how an ecospiritual praxis might help rethink the inter-relationality of our lives and of our schools based on ethical foundations of interdependence, justice, and ecological sustainability.

Notes

---

1. A "nexus" is a hypothetical node in a grid or a point where two or more lines or momentums intersect. It is a crossing where one of these touches another and forms a relationship that exists as it is only at that particular place and time. Alfred North Whitehead (1978) uses the term *nexus* to refer to an occasion of intersection between two or more "actual entities," i.e., anywhere the "actual real entities" from which the world is made come together in space and time—from a giant water oak to a red-headed woodpecker to "the most trivial puff of existence in far-off empty space" (p. 18). *Nexus* comes from the Latin *nectere*, meaning to tie, bind, or connect (*American Heritage Dictionary*, 1996, p. 1220).

2. My use of pronouns such as "we" and "us" is a general reference to larger questions of meaning pertaining to human beings as a species, and not meant to imply that meaning made from life experience will be perceived by any two individuals the same way.

3. Although each person's experience is unique to his/her circumstance, I would venture to say that within Western culture many are well acquainted with feelings of numbness, of separation, although the irony is that there are those who may be too numb to recognize it.

4. A mechanistic doctrine is a Newtonian view in which "the world is a vast machine of matter and motion obeying mathematical laws" (Berman, 1981, p. 42) and in which the human is seen as an alienated observer, a perspective on human consciousness derived from seventeenth-century science. Chapter 1 compares a mechanistic view with an organic one in which nature is thought to be "an enchanted world…wondrous [and] alive" (p. 42) in which the human is a *participant* in her/his surroundings.

5. Combining the terms ecology and spirituality, "ecospirituality" marks my understanding that ecology is an awareness of the interrelational nature of all that exists—to me, a spiritual awareness.

6. Cosmology is a view of the "physical universe considered as a totality…[including its] history, structure, and constituent dynamics" (*American Heritage Dictionary*, 1996, p. 424). Spretnak (1991) maintains that "[a]ll human experience and knowledge is situated in the unfolding manifestations of the universe, an interactive and genetically related community of beings" (p. 17). By *cosmological* world, I refer to the "integral reality" (Berry, 1988, p. 90) of all human and nonhuman forms as nested within a universal totality of relations. The interconnectedness of all matter is supported by atomic physics, according to Fritjof Capra (1975/1991), which views all "matter and

the basic phenomena involving them…[not] as isolated entities but…as integral parts of a unified whole" (p. 309).

7.   By ecospiritual, I refer to a perspective that views all planetary forms—both human and nonhuman—as being imbued with an infinitely creative dynamic "life force" connecting all things within an integrative web of relations.

8.   *Holism* is defined as a "theory that living matter or reality is made up of organic or unified wholes that are greater than the simple sum of their parts" (*American Heritage Dictionary*, 1996, p. 862) while *holistic* emphasizes the "importance of the whole and the interdependence of its parts" (p. 862). I use holistic to mean a recognition of the one and the many (James, 1907/1995), a part-to-whole relation (Capra, 1996; Zukav, 1989), a word which is derived from a common root—the Old English *hale*—which means health, heal, holy or hallow, suggesting a sacred dimension to the notion of the interconnectivity of all things within a healthy and interdependent ecological balance.

9.   In using the broad category "humanity" I refer to humans' larger identity as a species, one of a myriad of species inhabiting the earth.

10.   By social and biological matrix I refer to the integrative framework of mutually sustaining connections within which all life on the planet arises and has its being.

11.   Chapter 1 discusses the crisis of modern reason, when reduced to "rationalization processes" (Weber, 1968) such as reasonism, scientism, or corporatism, become instrumental means toward bureaucratic efficiency that override ends that are drawn for individual or social good.

12.   Discussed in Chapter 1.

13.   *Deep ecology* is a growing area within the environmental movement that "recognizes the fundamental interdependence of all phenomena and the fact that, as individuals and societies, we are all embedded in (and ultimately dependent on) the cyclical processes of nature" (Capra, 1996, p. 6). Generally, *ecofeminism* has been defined as a rubric under which people are working "to transform a social order that sanctions human oppression and environmental abuse" (Adams, 1993, p. 1). *Process theory* is a worldview that, unlike the rigidly deterministic view of Newton's mechanical world, "stresses the openness and indeterminism of nature" leading to an "organismic or ecological view of the universe" (Davies, 1992, pp. 182–183).

14. Chapter 1 goes more deeply into the subject of separation and how it has become one of the foundational characteristics underlying the notion of modern rationality.

15. A discussion of "consciousness" is beyond the scope of this book, yet it would be an interesting topic for further inquiry into "other ways of knowing." See Wexler, 1996, and also, Pinar, 1974 and 1999 for discussions of the notion of consciousness as having important implications for the field of curriculum theory.

16. By integrative being, I refer to a person being actively engaged within the wide range of sense-abilities that might be accessed at any moment in time (which, naturally, will vary), along with a recognition of her/his own vital connection within the social and ecological world. An integrative being is a *relational* being in all of its fullness—body, mind, spirit—engaged in an awareness of being *alive* within a complex network of relations.

17. My project draws principally on American Buddhism, my interest being primarily in what Buddhist principles have in common with perspectives of ecology and also with feminist theory. For further discussion, see Stephanie Kaza, 1993.

18. Drawing on postmodern and critical theory, any position may become totalizing when placed at the center.

Part One

[T]he origins of knowledge shape the way we see the world and ourselves as participants in it. They affect our definitions of ourselves, the way we interact with others, our public and private personae, our sense of control over life events, our views of teaching and learning, and our conceptions of morality.

—Belenky, et al., 1986

The Ground of Our Knowing

"Ways of knowing" in the twentieth-century West have been driven by a tendency toward separation, which fosters harmful anthropocentric and androcentric practices, relegating nature and women to the status of "resource," subordinated within patriarchal discourses. Part One explores ways that humans' epistemology has led to a crisis in ontology manifesting, in part, as a personal and cultural alienation affecting education.

Chapter One

Living Intentionally:
Process and Paradox

Quietly, I move through the screened doorway out onto the cypress deck, careful not to disrupt what's taking place in the early-morning world of suburban solitude. It's cloudy, late March, and the air feels damp, yet the clouds are clipping along too fast to expect that it will rain. The sun moves in and out of view and warms the air making a light sweater enough against the soft, gusting wind. Its fingers press my face and I watch it move the tender, greening shoots and leaves with cooler currents from the north. I step to the edge of the porch with my coffee cup, watching. A wood bee darts past and returns to hover near, inspecting my intrusion on his work. Every year these bees return to the wooden frame beneath this porch to bore their holes for laying eggs. Every year my landlord comes armed with an aerosol can in each hand doing combat with their spinning, darting ranks. They retreat with a few losses but still remain undaunted, cycling the season through to disappear with the late summer heat. I wonder how many sawdust-cycles before they take this place back to the elements from which every life evolves and toward which every death contributes.

Cycles are a part of living in this place. One comes to know the rhythms and routines of living process: that March will begin the yearly pilgrimage of large wood ants foraging along my counter-top for crumbs or droplets formed around the honey lid; that wind-blown tassels of oak and pecan will dust every corner of the world with a layer of powdery yellow pollen; that steamy showers of rain will wash every afternoon in August and that by the height of summer a lush canopy of sycamore will cut the sun's heat with a spread of cooling shade across my sloping metal roof. It's then that I'll stay inside siesta-ing in the artificial chill of air-conditioned comfort.

Life in Louisiana offers a luxuriant richness, slow, sensuous, and teeming with life. And it is marked by paradox. My home state is known for its haunting natural beauty as much as for contaminated streams and waterways; for its rich cultural heritage, food, and music, as well as a reputation for corrupt political affairs. In many ways Louisiana is appropriate ground for an inquiry into the complex ways we humans have come to view the world, as a "resource," giving priority to the human-use of the ecological world. It is a prime location to observe how we have come to *distance* ourselves from our own inner wisdom, from other people, and from our connection within the natural world.

It is this "distancing' perspective that I wish to address here. I would offer that the overarching crisis of our times is a crisis of *the way we think*. What has been a prominent current within the complex pool of Western thought for three hundred years is an underlying separation—of mind from body, from emotions, from spirit—that is fundamental to our "ways of knowing" (Belenky et al., 1986). The separation and reductionism woven through Western "rationality" underlies many of the socio-cultural difficulties troubling the West, including many problems within U.S. education. Often associated with what has come to be called Modernity,[1] justification of what is deemed "rational," has come to mean that which can be justified as instrumentally *useful* for those individuals holding power and making decisions (historically, white and male). I will explore the possibilities of expanding our[2] ways of knowing the world through a more ecological perspective based in relationality—drawing on ecofeminism, deep ecology, and process theory. My inquiries will circulate around problems within three areas: education, ecology, and what I perceive to be a vacuum within the human spirit. Many of the difficulties within all of these areas, I maintain, have a common basis in *separation*. Epistemological separation underlies a human-centered (anthropocentric) and male-centered (androcentric) model of "rationality" dominant within modern worldviews.

At least since Descartes' *Meditations*, in 1641, humans in the West have been trained to view the world in terms of separation. We divide "the thinking mind, the subject," from all else—from "the material world of things, or objects" (Abram, 1996, pp. 31–32), and also, even, from other aspects of our own internal modes of perception, "feelings, hunches, ways of recognizing complex patterns" (Furlong, 2000, p. 28), for example. The dominant rational mind objectifies, externalizes, makes all that is not "self-same" (Jardine, 2000) into an "Other," pushing it to the periphery of importance. As Fritjof Capra (1977) has said, "[C]ogito ergo sum—I think, therefore I exist—[has]...encouraged Western individuals to equate their identity with their rational mind rather than with their whole organism" (p. 377).Through Cartesian method, the rational mind has become a "subject" in juxtaposition to that which is not of-mind, an "object" existing as separate-from, indeed, Other. Descartes' subject/object split marked a beginning for Western "scientific method," which requires a distancing of the "subject" perceiving from the "object" in view, and then proceeds toward an end through a logical series of orderly steps (Berman, 1981): "subdivide, measure, combine; subdivide, measure, combine" (p. 34). Widely used throughout the natural and social sciences, the Cartesian method reduces a complex problem into its simplest form in order to perceive it "clearly and distinctly." The idea of *separation* is foundational to scientific method, wherein "truth" is "discovered" through objective method whereby humans may come to know the absolute truth of an object or process under study.

The Enlightenment era in Western Europe[3] is often constructed as a moment in history when the West made a turn toward the "scientistic reforming of society" (Spretnak, 1997, p. 85) and a flowering of the notion of human reason. The Enlightenment has been associated with ideals of a better quality of life for human beings with aspirations toward freedom and equality that could be realized through the vehicles of reason and science. Knowing has become synonymous with reason, the province of the rational mind. In Western culture, what we can know is considered to be provable through the "rational" view of objective observation. Reason can be defined in many ways. I will focus on two conceptions of reason/rationality:[4]

(1) Reason as an *analytical tool* that has given our mind a method to "confront the world as a separate object" (Berman, 1981, p. 34).

(2) Reason as a *social paradigm* rooted in the Enlightenment, out of which grew the Scientific Revolution with its surrounding social and technological changes.

The second arose as an embodiment of the first and, as some have argued, could be understood as a function of discourse, a linguistic phenomenon, a way of knowing situated in time and place and constructed within a context of cultural conventions (Munro, 1996).

The human ability to reason sets us apart from other species and has made it possible for science—and its corollary technology—to increase longevity and enhance quality of life in many ways. However, with scientific technology quality of life has also diminished. Subsequent shifts in practices of labor and production have increased industrialization and bureaucratization. The primacy of relationships, family, and community have been supplanted, at many levels, by an emphasis on materialism and competitive consumerism, leading to an increase of alienation among people. The abstracted view of "rationalism"[5] allows power-holding individuals and corporations to act chiefly with the motive of profit, exploiting planetary resources for personal gain. According to C. A. Bowers and D. G. Flinders (1990), this view is "based on assumptions that ignore the interdependence of cultural and natural systems" (p. 28). It suggests a lack of conscious awareness for life's sacred character and an irreverence for the sheer magnificence of the ecological system wherein we exist as *one* of a multitude of species. As G. Zukav (1989) tells us

> the cycles of life...have been in place for billions of years. They are the reflection
> of the natural breathing of the soul of Gaia itself, the Earth consciousness, as it
> moves its force fields and guides the cycles of life. If these are revered, how

could [humans] look at something as exquisite as our Earth's ecology and…risk the balance of this system? (p. 51)[6]

Linda Hogan (1996) maintains that "caretaking is the utmost spiritual and physical responsibility of our time, and [that] perhaps…stewardship is finally our place in the web of life, our work, the solution to the mystery of what we are" (p. 40). She uses the analogy of "tearing away" at the "fabric of life" (p. 40) to remind us of all of the life forms that have already been lost; all of the injustices that have been committed; all of the life we have not lived for we were too busy focusing on what we *have*, rather than on who we *are* (Fromm, 1986).

People in Western culture have been conditioned to gauge self-worth by material possessions and by job status rather than by the makeup of their inner being. Questions regarding the kind of person I am, what I value, how I approach my relationships with others and the natural world are diminished amid demands to compete and to succeed within a free-market economy. In twentieth-century United States, it is common for success to be measured by where you live, what you drive, and what you wear. Our technologically driven and capitalistic governing system condones, even encourages, the exploitation of natural resources on a scale which exceeds an equitable balance between human beings and the natural world. These practices can be linked with scientism, bureaucratization, and instrumentalization making up what Max Weber (1983) termed "rationalization processes,"[7] whereby "a type of rationality increases over time…[as, e.g.,] an increase in the efficiency of bureaucratic administration or the development of empirical science" (Bernstein, 1991, p. 53). These processes are intertwined within societal worldviews that drive "the technical utilization of scientific knowledge" (Weber, 1983, p. 28) toward the widespread domination of the natural world (Leiss, 1972), the sublated status of women, and the utilitarian development of educational systems within the United States.

Modern science has done much to improve the quality of human lives, yet beyond the many contributions of science, the "uses of science have [also] strengthened the most ecologically destructive tendencies of modern culture" (Bowers, 1997, p. 40). Science has been linked with personal and ecological destruction when it has been transformed into "scientism," as the application of science through an instrumental, means-ends orientation. This twist on the way that science has been used can also be said of reason, i.e., "reasonism," a world view based on the knowable, the certain, the absolute (Bernstein, 1991). Rainforests have been leveled, toxic chemicals released into the air, the soil, and the water, nuclear weapons possessing the power to destroy anything they touch for miles are targeted to annihilate highly populated metropolitan areas, all based on the "rationale" that to do so is acceptable and a necessary means to other ends—ends deemed more important than a responsibility for the planet. This drive toward destruction demonstrates a blindness to the reality of humans' connection within a biological matrix constituting the earth. Murray Bookchin (1991) has argued that

> our society has warped the best Enlightenment ideals, reducing reason to a harsh industrial rationalism focused on efficiency rather than an ethically inspired intellectuality; that it uses science to quantify the world and divide thought against feeling; that it uses technology to exploit nature, including *human* nature. (Bookchin, M. 1991, p. 59)

Thus, with reasonism and scientism, the trend becomes less a "utopian perspective" of "emancipation" through expanded knowledge and understanding, than an "increasing imprisonment of modern man in dehumanized systems" of bureaucratic normalization (p. 40). R. Bernstein notes how the notion of "purposive-rational" social action, also referred to as "instrumental rationality" (Weber, 1968), is driven by a means-ends orientation toward living at all levels and anticipates a great many critical theorists' charges[8] against utopian ideals of Enlightenment rationality (Bernstein, 1991, pp. 38–45). M. Bookchin (1982) reminds us of the dialectics of reason, a view acknowledging the paradoxical ways that rationality has played out over history. He explains that "the cherished concept of the Enlightenment—Reason," perceived as carrying the possibility for freedom and democracy and making possible human's unlimited perfectibility (pp. 33–35), has entailed, all the while, a desire to master reality through an "instrumentalization into technics" where reason has been deployed "as a tool or formal device for classification, analysis, and manipulation" (p. 269). He says that (1982), instrumental reason has failed

> to live up to its historic claim of emancipating humanity [and]...even...its more traditional claim of illuminating mind...[I]ts quest for innovation threatens to tear down the planet itself...verified by the foul air and water, the rising cancer rates, the automotive accidents, and the chemical wastelands, that assault the entire world of a scientistic "civilization." (p. 273)

Modern science, with its emphasis on "objective" understanding, has put high value on that which could be proven as "true," that which was undoubtable. The legacy of modern scientism—with its goal to understand, to know, to possess, to appropriate, to master, indeed to overcome ignorance—has become the modernist quest for the Holy Grail, setting up rationality as a deity or absolute through which all things may be known and thus attained. As a vehicle to help humans search for answers about "self" and "the world," reason has come to [be regarded] in such forms as

> civil law,...moral codes,...[and] the universal laws of humanity that claim to temper and prevent the violence that would supposedly exist without their civilizing constraints. (Dreyfus & Rabinow, 1982, p. 110)

Yet, as people of the planet begin the third millennium, the notion of human rationality has fallen into question because of what some consider to be the privileged place of reason within patriarchal conceptions of Western culture (Bernstein, 1991; Kohli, 1995; Pinar, 1996).

The Enlightenment has been equated with the Modern Age of Man[9] (Foucault 1973), which has privileged reason, "clear thinking," and a metaphysics of hierarchical structures (Bowers, 1987) based in separation. Separation is one of the primary characteristics of modern rationalism, an androcentric perspective. This white, male, Eurocentric (read patriarchal) view has delineated "legitimate" knowledge and espoused "the truth" for all people—even though a great many voices, a great many lives, a great many civilizations have been left out of the text (Munro, 1996). The current "rage against reason" (Bernstein, 1991) "evokes images of domination, oppression, repression, patriarchy, sterility, violence, totality, totalitarianism, and even terror" (p. 32). These images point to a contradiction in that the concept of reason has long been associated with notions of "autonomy, freedom, justice, equality, happiness, and peace" (p. 33).

Critical social theorists of the Frankfurt School were among the first to foresee contradictions such as these, most notably, I think of Horkheimer and Adorno's (1972) *Dialectic of Enlightenment*. Michel Foucault (1979) argues that any discourse, no matter how appealing, can become dangerous when held naively as unproblematic and thus beyond doubt (see also Bookchin, 1982; Haraway, 1991; Leiss, 1972). Trends countering the foundational structures of modernism such as reasonism, scientism, and individualism, e.g., have arisen out of particular places and times to give voice to the "unnatural" ways that modernity has produced us as beings embedded within normative structures and belief systems (Spretnak, 1997, p. 2). Effects of these rationalization processes (Weber, 1968) have left human beings with a sense of alienation and disconnection from themselves, from their communities, and from the natural environment. For my purposes, I will suspend discussion of the broader social phenomenon that has been referred to as the "Age of Modernity" in order to narrow the focus to a discussion of alienating effects rising out of "modernism," a culturally constructed constellation of societal beliefs and practices that have been institutionalized into twentieth-century Western society.

The tangled web of characteristics often associated with the institutionalized modes of knowing and being that have come to be constructed as "modernism" deny any pretext that it could represent a pure form. Full of complex and sometimes contradictory trends, modernism has been referred to as a "conjoining of the ephemeral and fleeting," with the "eternal and immutable" (Harvey, 1989, p. 10) in a perpetual state of tension. We have been conditioned to view history as a linear stream wherein periods are constructed as one coming on the heels of the last, with beginnings and endings demarcated and labeled as eras. The premodern, modern, and postmodern ages have come to be thought of in this way, along with their corollary constructions of thought. There is overlap and contradiction

throughout, however. Historical eras represent a complex maze of positionings that are more a confluence of commingling and countervalent
forces than clear-cut divisions in linear time. There have been many who
have critiqued the foundationalist tenets of modernism from various
standpoints, even while existing side by side with or as components *within*
modernism. According to Spretnak (1997), there have always been anti-
foundationalist tendencies existing within modernist trends such as

> the loss of faith in scientific positivism that began in the late nineteenth century,
> the cantankerous perspectivalism of Nietzsche, the...[idea] of "language games"
> illuminated by Wittgenstein, the sociology of knowledge, and the various
> political critiques of rationalist, patriarchal, racist framings of reality that were
> put forth by grassroots movements of the 1960s. (p. 67)

I am drawn to more recent anti-foundationalist critiques (Deleuze,
1977/1987; Doll, 1993; Capra, 1996; Gergen, 1991; Hayles, 1991) from
areas of postmodernism in attempting to disrupt what I see as some of the
destructive characteristics of modern philosophical thought. I do not intend
to reinstate another binary by posing modernism against postmodernism;
however, I would like to draw on some of the theoretical strategies that
postmodernism offers for questioning dualism, such as, thinking in binaries
of either-or that is the basis for anthropocentric and androcentric worldviews. Any theoretical base[10] from which we draw—modernism, postmodernism, ecofeminism, for example—consists of foundations upon
which ideas rest. These "conceptual frameworks" (Warren, 1993, p. 122)
use categories to inscribe meaning. Even those positions that claim *anti-*
foundationalism enlist the uses of foundations that allow humans to make
sense of the world. That "truth" does not exist may be "truth" to a
postmodernist. Theoretical categories provide the means to communicate
certain ideas, while simultaneously preventing other conversations. In this
way, the language in which we live, think, and write carries with it
possibilities and also limitations. Whether one sees the figure or the ground
reflects how one views the world. Philosophical orientations are culturally
constructed and define and delimit the ways humans come to know.
Theories seldom exist within lived experience in the pure form, as they

might appear on paper. People use them for specific purposes, selecting, rejecting, testing, combining them with others, toward the construction of meaning. Recognizing that there are possibilities and limitations within any theoretical base, I propose to move within the margins *between* theories, where I might draw from a plurality of differing theoretical regions: postmodernism, ecofeminism, deep ecology, even modernism. To do this, I would like to engage in a postmodern movement of "multiple layering" in order to blend the many disparate forms in not-always-likely combinations. This weaving of forms allows investigation into some of the underlying theoretical structures of modernism that have led to the problems of alienation within and among human beings, and also between humans and their environment. It will simultaneously allow me to retain elements of modernism that I construe as being valuable—inquiring into what modernity has shown us, learning from what has been, in order to move toward constructing a different reality from that which we have known and found lacking.

Postmodern Perspectives

The name *post*modern suggests a moving-beyond the search for "truths" or "certainty" or the "authentic" nature of what *is*, preoccupying modernist philosophy at least since the Enlightenment 1600s, and even going back to the Greeks. Postmodernism cannot be defined as a single unified philosophy based on a foundation of common beliefs, since it constitutes a wide array of varied theoretical positionings which sometimes contradict one another. However, there are some common anti-foundationalist tendencies that can be noted across the varied strands of postmodern thinking. Most postmodernists are suspicious of any totalizing Theory of Everything[11] (Davies, 1992). They believe that grand narratives about "how the world is" (Spretnak, 1997) have been used oftentimes through history to 'rationalize' corruption or oppression or destruction of human life (Kohli, 1995). For many postmodernists, it is a dangerous practice to overly determine any position into the center as "the truth," for fear of risking the position's totalization. Postmodern worldviews are characteristically marked by multiplicity, fragmentation, contingency, and change and question the taken-for-granted ideas of "deep meaning," or "commitment to principles," or "metanarratives" (Gergen, 1991; Lyotard, 1979), including the assumptions underlying postmodern theory itself.

One idea with which many postmodernists disagree is the notion that the truth that we can know is independent of the knower (Munro, 1998, p. 5), an *a priori*, predetermined form to be discovered or revealed rather than created in the moment-by-moment of living experience. Many modernists deny ways that truth and knowledge, indeed meanings, are negotiated between unique individuals and that there are varying and contradictory subjectivities existing within one individual, their formation influenced in various and specific ways, for example, by gender or race or class (Bloom & Munro, 1996). Postmodern theory has taught me that the "individual" is no longer a "unified self" but is constituted through language and represents multiple "positions" or "interests" (Kohli, 1991a, p. 58), so that the "inner self" I wish to know is more a confluence of forces, never the same unitary being capable of speaking authentically, or of coming to "the Truth," at last. Rather than a modernist idea of the self as being an "autonomous individual," the postmodern self is relational, replaced by the postmodern *subject*—seen as a constantly shifting, changing form, more an "assemblage" (Deleuze, 1977/1987) than a single, unified individual. Feminist poststructuralist Wendy Kohli (1991b) notes that the "[t]he de-centered subject' is now a code word for the postmodern world" and that from many postmodern perspectives "each of us exists within and through the expression of 'multiple subjectivities,' subjectivities that are constituted through language" (p. 39). Petra Munro (1998) maintains that "[n]otions of the self as unitary, autonomous, universal and static are fictions" (p. 34). It follows, then, that modernist discourses, such as "individualism," are in question.

Within a modernist perspective, linked with capitalist democracy, value is often placed on the "self" according to its ability to reason and to compete successfully within a free-market economy. C. Spretnak (1997) points out that "the human is considered essentially an economic being, homo economicus" (p. 40), driven by a desire for consumption of material goods and services pervasive within Western culture. The Westerner is ideally "self-reliant, self-motivated, an individual who is self-directing" (Gergen, 1991, p. 240), a notion resonant with the purposive directions of capitalism, since the individual as a "decision maker…has a right to buy and sell…[toward] the common good" (p. 240). The modernist valorization of the subject as a self-directed maker of decisions, or a creator of the person s/he is to become (Gilmore, 1994) dominates mainstream ideals in the West, so that the idea of the "one" has occluded the "many" from view.

   Along with the emphasis on the "individual" as the primary unit of our culture (Bowers, 1987) another idea associated with the theoretical construct of modern thought is the Western ideal of "progress." In other words, there is a widely held (modernist) assumption that people move through life on a forward trajectory of expansion and growth, and this is seen as progressive. Coupled with this idea of escalating growth, the individual's right to buy and sell in a capitalistic system has placed "economic expansion [economism] and technological innovation [technocracy] at the center of importance" (Spretnak, 1997, p. 2). Bowers (1993) questions the

> cultural assumptions underlying the belief systems of…developed countries whose technologies and patterns of consumer-oriented living are depleting the world's energy resources at an alarming rate. The core values of this belief system—abstract rational thought, efficiency, individualism, profits—were at one time believed to be the wellspring of individual and social progress." (p. 3)

Material wealth and social standing have come to be linked with the idea of social progress leading to a common belief in developed countries that socio-economic status should increase beyond the level of one's parents. Consumerism continues to escalate at a rate in excess of the earth's ability to sustain energy consumption. Yet, progress has become a cultural "given." Humans take comfort in the idea of progress. When linked with the modernist tenet of "rationality," it allays our deepest fears of what we do not know and cannot see because implicit in the modern idea of "progress" is the notion that we move away from unknowns toward a state of certainty and control (Spretnak, 1997).

Modern rationality often relies on the separation of phenomena into two opposing principles or dualisms, for example, mind or body, good or evil. These black-or-white judgments of the truth of a matter often prevent our being able to see all that lies *between* two fixed points of reference. Certain postmodernist/feminists and feminist/epistemologists have offered valuable criticisms of ways reason/rationality and the Enlightenment Project are bound up with many contradictory elements. These dualisms take many forms, but one of the most long-standing may be the patriarchal power relation held by "man" over "woman" throughout history, an overt and long-running manifestation of social domination legitimated by reason. "Patriarchy" is the systematic dominance of men in society" (King, 1990, p. 109) based on dualistic and hierarchical ways of conceiving one's relations with the world. In terms of men and women, this relation has manifested in patterns of male-subjugating-female, rationalized as "natural" since the Garden of Eden. Its subsequent legacy of abuse and struggle culminated in the growth of the contemporary feminist movement, arising in the West in the late1960s and early 1970s.

Contemporary feminist critique is positioned ambiguously within postmodern debates concerning the foundations of cultural thought. While there are postmodernists and feminists sharing some views, "modernist values are very much a part of contemporary feminist positions," rooted as the latter often are in the "emancipatory impulse of [either] liberal-humanism...[or] Marxism" (Hekman, 1990, p. 2). The very issues which people defend, politically—truth, justice, ethics—are subject to question within some postmodern discourses. Susan J. Hekman (1990) explains that the

> contradiction between...[modernist] values and the postmodern themes of much
> of contemporary feminism thwarts attempts neatly to categorize feminism as
> modernist or postmodernist. (p .2)

Feminist epistemologists (Harding, 1991; Jaggar, 1989; Keller, 1985; Longino, 1989) have been preeminent in asserting that "the model of knowledge embodied in the scientific method...is not the only paradigm of knowledge" (Hekman, 1990, p. 4). They have also been part of anti-foundationalist challenges to modernism for "the dualisms on which Enlightenment thought rests," exemplified in linguistic binaries, such as subject versus object, reason versus emotion, and culture versus nature (p. 5). There are feminists who assert that these dualisms

> are a product of the fundamental, dualism between male and female…[wherein] the male is associated with the first element, the female with the second. And in each case the male element is privileged over the female. (Hekman, 1990, p. 5)

Ecofeminism has been valuable for exposing androcentric practices at the root of "both social hierarchy and the destruction of nature" (Dingler, 1999, p. 2). Susan Griffin (1990) speaks of our "fragmentary vision expressed in the categories of masculine and feminine…not the biological male and female," but the gendered categories "masculine" and "feminine" that have been socially created (p. 87). The category "female" has been constructed through modern epistemologies as "object" to the dominant male "subject," as has nature, both construed as being "culturally passive and subordinate" (Merchant, 1980, p. xvi). Hierarchical ways of knowing, grounded in the dualisms of patriarchal thinking, have kept both women and the natural world in positions of subordination for much of history.

Another binary within patriarchal ways of knowing that is foundational to Enlightenment rationality is the dualistic split between culture and nature. Carolyn Merchant (1980) has said that the concept "nature" has meant different things to different people over the course of history. In ancient times nature was associated with certain "properties, inherent characters, and vital powers of persons, animals,…things, or…generally to human nature" (p. xxiii). Nature was construed as an "impulse" coming from within which caused one to act or…in resisting…action," one was said to "go against nature" (p. xxiii). Nature has also been perceived as female, and the "course of nature and the laws of nature…[have been perceived as] the actualization of her force" (p. xxiii). Max Horkheimer (1947) insists that "the disease of reason is that reason was born from man's urge to dominate nature" (as cited in Leiss, 1972, p. 148). Furthermore, the collective madness that ranges today, from the concentration camps to the seemingly most harmless mass-culture reactions, was already present in germ in primitive objectivization, in the first man's calculating contemplation of the world as a prey. (p. 148)

With the Scientific Revolution humans increased their abilities to use "science and technology as instruments designed for the conquest of nature" (Leiss, 1972, pp. 101–102). William Liess (1972) has said that an instrumental approach can be seen underlying this conquest. He draws on Max Scheler's (1960) discussion of *herrshaftswissen*, described as "knowledge for the sake of domination," in exploring the relationship between the domination of nature and the development of the sciences (as cited in Leiss, 1972, p. 105). Scheler asserts that "the conceptual structure of modern science is 'designed' for the mastery of nature" (p. 115). "Knowing" has been linked historically to the concept of domination, which is implicit in humans' struggles against the dangers and uncertainties of the natural environment (Leiss, 1972, pp. 105–106).

In ancient times, large numbers of people commonly turned to religions in order to reassure themselves in an uncertain world where they were forced to struggle against nature in order to maintain their existence (p. 106). Humans' desire to gain a sense of control over their lives caused them to pay homage to the 'spirits' that inhabit all aspects of the natural world, attempting to ensure themselves "against harm…placat[ing] the spirits through gifts and ceremonies" (p. 30). As Christianity supplanted animal paganism, a shift was made. Humans could justify exploiting nature due to Christian beliefs that human beings were separate from and above the rest of the natural world (Leiss,1972). Christian traditions supported the view that all resources were put on the Earth for human use. It was the human birthright, according to Genesis, to master "the fish of the sea, the birds of the sky, and all the living things that creep on the earth" (Abram, 1996, p. 94), a seeming justification for the domination of a natural world "designed to serve man's [sic] ends exclusively" (Leiss, 1972, p. 30). Carolyn Merchant (1980) concurs that humans have historically used religion to rationalize self-serving, anthropocentric uses of nature said to be the will of God. At the base of the Judaeo-Christian perspective was a principle of separation that "maintained that 'spirit' was separate from nature and ruled over it from without" (Leiss, 1972, p. 30). In this separation, many humans have seen themselves as standing

> apart from nature and rightfully exercis[ing] a kind of authority over the natural world...a prominent feature of the doctrine that has dominated the ethical consciousness of Western civilization. (p. 32)

With the Modern Age, science came to take the place of religion for many people looking for security in an uncertain world. One of the most influential advocates for the development of science and technology (p. 47) was Francis Bacon, who viewed "the conquest of nature" as a promise of

> liberation from the...adverse conditions of existence which arise out of the prevailing state of the relations between [humans]...and nature. (cited in Leiss, 1972, p. 56)

Bacon used language which cast the concept of nature into pejorative feminine terms "displaying strong overtones of aggression" (p. 60). Merchant (1980) says that nature has long been identified as female, both as a nurturing mother and also as being "wild and uncontrollable...render[ing] violence, storms, droughts, and general chaos" (p. 2). The former image of nature as "a kindly beneficent female who provided for the needs of mankind in an ordered, planned universe" (p. 2) began to diminish with the Scientific Revolution, while the latter image of "nature as disorder, called forth...the modern idea...of power over nature" (p. 2). Merchant (1996) notes that Bacon's "description of nature and his metaphorical style . . .were instrumental" toward the subsequent shift in European thought, "which transform[ed]...the earth [from] a nurturing mother and womb of life into a source of secrets to be extracted for economic advance" (p. 80).

Another dualism framing many modernist viewpoints is that of the earth as an organism versus a mechanism. An idea associated with nature in the ancient world was that of "organicism," a concept used to name the interrelational workings of nature, society, and the cosmos. Physicist David

Bohm (1985) traces an organic view back to the "ancient Greek notion of the earth at the center of the universe" as part of an integral organism having "activities regarded as meaningful" and interrelated (p. 1). The two female images of nature, the nurturing mother figure and that of a female tempest, were a part of organic theory in ancient times; however, the former seemed to vanish from prominence as the Scientific Revolution came to the forefront. Nature as a turbulent female-to-be-tamed and dominated came to be accepted and has wide metaphorical use within Western patriarchal culture. According to Merchant,

> [a]n organically oriented mentality in which female principles played an important role was undermined and replaced by a mechanically oriented reality that either eliminated or used female principles in an exploitative manner. (p. 2)

Mechanistic worldviews can be traced, in part, to Isaac Newton, who was influential in laying a "foundation for eighteenth-century experimental philosophers who wished to…reduc[e] known phenomena to simple laws" (Merchant, 1980, pp. 278–279). Newton described "material particles" as "rearrangeable into new configurations by the actions of external forces" (p. 278). The idea that passive matter was acted upon by external stimuli denied the internal initiative implied within organic theory and "provid[ed] a subtle sanctioning for the domination and manipulation of nature necessary to progressive economic development" (p. 279). The shift from organicism to mechanism also came to replace the female earth spirit with that of a machine. Bohm (1985) contrasts an "organic" worldview beside one he describes as "mechanistic." The latter he says has "obtained its most complete development" in the world of physics and has spread to "almost all fields of human endeavor" (p. 2), permeating the way we tend to look at life.

An important difference between a mechanistic view and an organic one is in the perception of the relationship between parts. In an organic view the

very nature of any part may be profoundly affected by changes of activity in
other parts, and by the general state of the whole, and so the parts are basically
internally related to each other as well as to the whole. (p. 3)

A mechanistic view "does not constitute a whole with meaning. . . [rather]
its basic order is that of independently existent parts interacting blindly
through forces that they exert on each other" (pp. 1–2) suggesting that they
have only an external relation to one another. In the Newtonian framework,
"the world is reduced as far as possible to a set of basic elements...with the
fundamental nature of each [seen as] independent of that of the other," a
mechanistic perspective. Bohm explains that these elements aren't thought
to "grow as parts of a whole, but rather...influence each other externally,
for example, by forces of interaction that do not deeply affect their inner
natures" (pp. 2–3).

Postmodern thinking has generated pathways that move human ways
of knowing beyond those framed in reason-ism, scient-ism, material-ism,
instrumental-ism, and dual-ism. The modernist search for the *is* has been
replaced within postmodern circles by a foregrounding of multiplicities and
relations. French poststructuralist Gilles Deleuze (1977/1987) has said that
the "history of [Western] philosophy is encumbered with the problem of
being, IS" (p. 56), and suggests that what must be done is to

make the encounter with relations penetrate and corrupt everything, undermine
being, make it topple over. Substitute the AND for IS. A and B. The AND is not
even a specific relation or conjunction, it is that which subtends all relations, the
path of all relations, which makes relations shoot outside their terms and outside
the set of their terms, and outside everything which could be determined as
Being, One, or Whole...the AND gives relations another direction. . . [a] line of
flight which it actively creates (p. 57).

There is much I am learning from postmodernism. It is becoming clear
that a perspective frames a person's view of the world, delineating possible
"ways of knowing." Worldviews are constructed through life experiences
and inscribed through the stories an individual chooses to tell—both to self
and to others—and are shaped, largely, by what is included and also by
what stories are never told. Thus, the stories I tell are my own, constructed
from my context and brought into form "through the languages of the

public worlds, constituted by [the] disciplines and...the institutions within which the disciplines are organized" (Pagano, 1990, p. 2). This is also a story of my own becoming, my own ways of "gaining some facility with the conventions of the narratives that structure...disciplines and institutions" (Pagano, 1990, p. 2). "Discourses in any field," Pagano says, "define the stories that can be expressed; they permit certain stories to unfold, and they forbid others" (p. 2).

The field in which my interests lie is education. My background as a white, middle-class female raised within the twentieth-century United States has shaped my view toward education through a lens of certain liberal-democratic values and assumptions. I have grown up with the idea that there are "truths" by which we live our lives, and ideals toward which we may strive in becoming a better person; that we make choices about what is "right" or "wrong" based on reasons; that what is "right" leads each one toward "the good;" and that what is "good" for the individual will lead toward what is "good" for all of society. These "first principles" are not mine alone but exemplify some of the *cultural* beliefs and "truisms" to which I have subscribed, and they are based on many things, including my gendered, raced, and classed "positionings" (Kohli, 1995) I have become aware of some of my own assumptions underlying these "first principles." Among them is the idea that what is "the good" for one, if not the same, will be at least compatible with what is "the good" for other individuals, and with society as a whole. I wonder.

As I lie here fresh from sleep, watching a spider spinning gossamer trails through a thicket of branches just outside the window screen, I am aware that the web glistening in the February sunrise will soon snare his morning meal. I am coming to see that how one views the world depends *so much* on one's own personal lens, the perspective that has been formed through all of the impressions we've had and experiences we've lived during the course of our lives. And it has been shaped through the historical, cultural, and linguistic systems within which we've come into being in particular ways, each person being unique within specific contexts of time and place, the nexus of relations. Perhaps what could be called a new era *is* upon the world. I am beginning to understand the ways that many of the modern precepts that Western tradition held as true, noble ideals since the Enlightenment, have also been used to silence and exclude and even prey upon those who were unable or unwilling to defend their positions.

I admit there was a time when I held to the idea of a unitary subject and that there were universal truths by which we could know, finally, all things at once, a "Theory of Everything" (Davies, 1992, p. 21) uniting every field. If reality existed, which all my senses verified, then there had to be some vision wide enough to include it all under its vast expanse, a grand metanarrative explaining what *is* under its subsuming reach. I am coming to terms with challenges to my own normative assumptions (Kohli, 1995), challenges that tell me that universal or essentialist ideas such as "truth" or "autonomy" are problematic (Lyotard, 1979).

Postmodernism has been valuable for disrupting taken-for-granted assumptions on which we base worldviews. "Whenever one believes in a great first principle," Deleuze (1977/1987) has said,

> one can no longer produce anything but huge sterile dualisms. Philosophers willingly surrender themselves to this and center their discussions on what should be the first principle (Being, the Ego, the Sensible...). But it is not really worth invoking the concrete richness of the sensible if it is only to make it into an abstract principle. (p. 54)

He notes that the "first principle is always a mask, a simple image" (p. 54) and says that things do not really begin to "come alive until the level of the second, third, fourth principle, and these are no longer even principles. Things do not begin to live except in the middle" (pp. 54–55).

I admit that, at times, I find myself embroiled in dualistic thought, some position of either/or which implies that there-is-black or there-is-white and which clouds my view from much which could lie in the *between*. Linguistically, as a preposition, the "between" is a positioning of relationality. Vikki Hillis (personal communication, 2–18–99) draws from Jacques Derrida (1993) in terming it the "aporia," that passage beyond brick walls of dualism or incommensurability (Lyotard, 1979), that site wherein the wall becomes a passage, a moving-through. It's in the prepositions, Hillis says, the context wherein relationality may allow for a transcendence of the impasse, where incommensurability may be moved "through" and moved "beyond".

It is those between-spaces I'd like to evoke in considering how we have come to where we are and how we might shift our perspective toward somewhere in-between the modern and postmodern, between certainty and question. We never arrive finally at that which is "Truth" because every question gives rise to another and what is "true" and "real" is so only at the nexus of relations that exist within a particular context. Postmodernism has demonstrated that our categories such as "Truth" or the "autonomous individual" no longer serve us. Could there be a world beyond categories, or totalizing discourses? Maybe not. But could we reorient our view past perspectives that narrow the range of choices into dualisms, frames of either/or? Perhaps the in-between spaces may offer ways of articulating *across* discourses of difference, recognizing that one and "the Other" each have unique backgrounds and histories, each contingently situated within the nexus of relations.

Postmodern theories allow me to draw from a variety of traditions, weaving together that which lends richness to the many-layered vision I construct. Postmodern discourses of multiplicity, particularity, and difference make it possible to critique foundations of modernism, while holding on to certain first principles that seem to me to be valuable and real. In this regard, I am neither a modernist nor postmodernist, but choose to construct my view on a ground somewhere between the two. I propose an ecological postmodernism that is willing to let go of a metanarrative, replaced by narratives-within-narratives, a grand narrative, as William Doll (1993) says, with no final frame but which is open and generative of further perspectives, wider views without end. The parameters of such a worlds-within-worlds frame are never closed but always "aspir[ing] to open up rather than close off possibility" (Rasberry, 2001, p. 24) to new arrangements, new combinations, a dynamic form.

In terms of the "individual," I am willing to let go of the idea of privileging the "one" and, instead, balancing it with a notion of the "many" as equally vital, a part-to-whole relation. Drawing on conversations with

Zen Master Shunryu Suzuki, David Chadwick (1999) points to the shifting nature of the part-to-whole relation as "the duality of oneness, the oneness of duality" (p. 346).[12] From a different angle, pragmatist philosopher William James (1907/1995) speaks of the debate between monists and pluralists as "the most central of all philosophic problems" (p. 50), in pointing to problems with "unity," which have subsequently come to underlie postmodern critiques of the individual self. Noting that philosophy "has often been defined as the quest or the vision of the world's unity," James questions our strong emphasis on the *one* while tending to belittle the importance of the *"variety* in things" (p. 50). He does not dichotomize the one and the many because his "many" is inclusive of the one, while the "one" he describes is composed, also, of the many. He quarrels with a "certain emotional response to the character of oneness, as if it were a feature of the world not coordinate with its manyness, but vastly more excellent and imminent" (p. 50). James suggests that "acquaintance with reality's diversities is as important as understanding their connexion [sic]" (p. 50). It appears to me that balance among the varied positions is preferable and in line with the Zen notion of the "middle way," where it is recognized that one can't "speak the whole truth, there [is] always another side created by whatever [is] said" (Chadwick, 1999, p. 346). In 1979, Lyotard asks that we "wage war on totality...be witnesses to the unpresentable...activat[e] the differences" (p. 82), a war[13] which has been waged under the banner of the postmodern. This has opened up a multivocality, allowing for contradiction, critique, a cacophony of Others (Pinar et al., 1995, p. xiii) to be heard in all the range of their alterities.

Notes

1. Modernity could be described as a "condition [which] grew out of the Renaissance until, in the nineteenth century, it gave birth to cultural modernism" (Jencks, 1992, p. 6).

2. By using pronouns such as "our," "we," or "us," I do not mean to stereotype in ways suggestive that *all* human beings can be essentialized into broad categories. Extending particular positions as if they apply across all people tends to overlook the ways we are each unique beings and constructed differentially through culture, background, gender, etc. When I speak in general terms, I direct my comments to issues that affect the earth and its people as a species, and also to the broader existential level in the spirit of offering to people still another point of view to be considered.

3. Beginning around the late 1600s.

4. I use the terms reason and rationality synonymously, unless otherwise indicated.

5. Seventeenth-century Western philosophy has taught us to think "rationally" through a distancing perspective so to objectively gauge that which is of value from that which is not. The alienating rationale of *objectivity* has been used by people in power (usually white and male) to "denigrate, suppress, or marginalize" (Bernstein, 1991, p. 58) those people or positions that would stand in the way of whatever means would accomplish their self-serving ends—in this case, ends deemed more important than a responsibility for the earth.

6. The concept of the earth as an organic body, "Gaia," is discussed in chapter 3. [For further reading see James Lovelock's (1979) *Gaia: A New Look at Life on Earth*. New York: Oxford University Press.]

7. The expansion of the rational model as institutionalized within bureaucratic systems for management and control.

8. See, for example, Horkheimer & Adorno, 1972.

9. Foucault's term—beginning in the 1700s—to designate the period in history when human grows to be "a special kind of total subject and total object of his own knowledge" (as cited in Dreyfus & Rabinow, 1982, p. 18).

10. By "theoretical base" I mean a conceptual framework as "the set of basic beliefs, values, attitudes, and assumptions that shape and reflect how one views oneself and one's world,... socially constructed lenses or filters through which one sees oneself and others" (Warren, 1993, p. 122).

11. Paul Davies (1992) uses the term "Theory of Everything" (TOE) to refer to a "complete description of the world in terms of a closed system of logical truths" (p. 21).

12. David Chadwick's biography of Suzuki—author of the Zen classic, *Zen Mind, Beginner's Mind*—places the Japanese monk as being the founder of the San Francisco Zen Center and of *Tassajara*, the first Zen monastery in the West located ten miles inland from Big Sur, California.

13. "War" is an unfortunate metaphor. I'd prefer, perhaps, a symphony of harmonic dissonance.

Chapter Two

The abstract intelligence...is a sheer delight when it's in service to the earthly dance, but reckless and stiflingly mean when it strives to certify its dominion...stand[ing] forever outside the sensuous world.

—David Abram, 1996

Other Ways of Knowing as Ecospiritual Praxis

My Louisiana home is a small frame house dubbed "the cabin" by a legion of graduate students passing it down by word-of-mouth over most of two decades. Less a cabin, than a tidy little "shack," it is nestled in a swirl of pecan, sycamore, and elm along the old Highland Road, historic trail of high ground settled two hundred years ago by Spanish, French, and Dutch plantation builders. Today, their historic legacies of Greek revival columns and vine-covered *portecocheres* span this winding road from the university gates going ten miles to the south. These dwindling estates hold their silent vigil against encroaching gas stations, cappuccino stands, and upscale suburbs with names like Highland Bluffs, Majestic Oaks, and Plantation Ridge. The Highland Road marks a western parameter of town above the backwater lowland of the Mississippi River. Lined with overhanging oak, dogwood, and telephone wire, the now-bustling, asphalt two-lane provides an alternate route for outlying commuters coming into town and choosing its aesthetic charm over the parallel Interstate 10.

Humans perceive a lifetime of potential "alternate routes" comprising every cross-road decision we come to. But how much do we really choose the courses we will follow and to what extent are they shaped by the linguistic/historical/cultural grid within which we're a part (Bowers, 1987; Foucault, 1979)? How do we come to know what we know? How much are our actions born of our choosing, and how much are they a function of the "ways of knowing" (Greene, 1995) society has sanctioned? And so I'm thinking about knowing and what it is I mean when I speak of *relational* ways of knowing. What is the vein of "knowing-ness" that I wish to explore here? Susie Gablik (1991) has observed that our culture "works by legitimizing certain ways of knowing and disqualifying others" (p. 46). Reason has come to be a paradigm that has privileged particular points of view as having "the truth" because of claims to "clear vision," while penalizing, either straightforwardly or by omission, those who base their lives on more intuitive or subjectively contextualized positionings. William Pinar (1996) has noted that "[r]eason is the regime in which and through which, our voices are raised, the medium through which we are coded as intelligible or not" (p. 10). Unfortunately, Western ideals of rationality

> distort and leave partial our understandings of nature and social relations by
> devaluing contextual modes of thought and emotional components of reason.
> (Harding, 1991, p. 118)

As mentioned earlier, the paradigm of scientific investigation turns on the idea of objectivity, presuming that the researcher is able to "stand outside" and distance her/himself from the sway of subjective forces such as emotions, which could bias the findings and thus contaminate the study. In this regard, modern theories of knowledge lead us to surmise that the scientist is the "dispassionate investigator" who is able to leave emotions out of the investigating process; however, some feminist theorists (Griffin, 1990; Harding, 1991; Jaggar 1989; Keller, 1985; Kohli, 1984b) suggest that such a supposition is unrealistic or even impossible. For example, Alison Jaggar (1989) asserts that "[o]bservation is not simply a passive process of absorbing impressions or recording stimuli" but rather is an "activity of selection and interpretation" and that what is selected for investigation and "how it is interpreted are influenced by emotional attitudes"(p. 154). Reason and emotion are interactive processes. Emotions shape "objective" reason by "focusing our attention selectively, directing, shaping, and even partially defining our observations" and in turn, "our observations direct, shape, and partially define our emotions" (p. 154). This interactive process between reason and emotion, Jaggar (1989) says,

> suggest[s] that certain emotional attitudes are involved on a deep level in all observation, in the intersubjectively verified and so supposedly dispassionate observations of science as well as in the common perceptions of daily life. (p. 154)

Just because individuals are often unaware of emotions, does not mean that emotions are not present under the surface, influencing the ways we articulate values, observations, thoughts, and actions (p. 156). Jaggar suggests that values inform the decisions that scientists make as to what to investigate, how to go about an investigation, and that they play a role in the interpretation of the results leading to a solution. I agree with Jaggar that there is a need to "rethink the relation between knowledge and emotion" (p. 156) rather than repressing emotion as an unimportant factor in scientific inquiry. I would suggest an approach that emanates from somewhere *in-between* reason and emotion, creating "conceptual models that demonstrate the mutually constitutive rather than oppositional relation between reason and emotion" (p. 157). Dualistic thinking, artificially imposed by patriarchal models of scientific investigation, categorizes concepts like reason and emotion into binaries, reducing the research process into simplistic pre-given categories. Griffin (1990) argues that in "[u]sing scientific method, scientists attempt to be above sensual experience. But instead of being above experience, they are perceiving partially" (p. 87). This is a perspective on research that limits the directions that inquiry may take and narrows potentials it may have for more thorough understandings of particular problems under study.

Gregory Bateson (1979) has said the heart has its reasons which reason cannot name (as cited in Berman, 1981, p. 197). Although trained in the sciences, Berman says that Bateson defended what was "not attainable by rational calculatio," and was strongly influenced by the ideas of his biologist father, Maxwell Bateson, who posed "intuitive insight as evidence for the view that there was a limit to the truth of any scientific explanation,...a deeper level of reality which lay beyond its reach" (p. 201). There is much about living life that lies beneath our abilities to experience it with merely the intellect. Maxine Greene (1995) maintains that

> rationality itself is grounded in something prerational, prereflective—perhaps in a primordial, perceived landscape…The conditions of objectivity, of course, have to do with the vantage points of the embodied consciousness, moving, seeing, touching, hearing in the midst of things…The preflective, that is, what we perceive before we reflect upon it, becomes the launching place of rationality. (p. 53)

But how to extend legitimation, also, to that "prereflective" zone—that intangible, yet palpable, realm of our corporeal existence that is so much a part of the substance and ground of our knowing? I heard it said recently, that "nothing enters the mind except through the gateway of the senses" (LaMothe, public address, Unitarian Universalist Church, August 1999). From observing myself, it appears to me that this is so. Simultaneously, the sensate experience of the body, the emotions, the spirit are given meaning through the mind's processes of interpretation. It is the mind *with* the sensorial context—figure *and* ground—that colors and deepens the substance of day-to-day life, including the lives of those in schools. Yet, we attempt to insert a wall between mind and body, creating an illusion of division from much that is real and concrete through the abstract objectification and the reification of the mental, rational domain of human experience (Spretnak, 1997). Focused, as Jardine (2001) tells us, on "Cartesian dreams of an earth full of isolated …'objects' bereft of rela-tions," many humans lose sight of the "fleshly, earthy" (p. 274) sphere of a world "full of smells and names and faces and kin, full of ancestral roots and ongoing conversations and old wisdoms and new, fresh deliberateness and audacity and life" (p. 278; In Hocking, et al.). We overlook so much of what makes learning and living synonymous—learning *as* living— what makes it generative, that freshens and renews.

Hocking, Haskell, and Linds (2001) speak of a type of "embodied learning that might be experienced and revitalized at all levels" of the educating process. They combine the words *body* and *mind* without hyphen, "bodymind," to symbolize a desire to move "beyond…an atomistic and binary view of the human being" (p. xvii). Bodymind is a term that describes learning as *lived experience*, education as a "way of being through which our embodied awareness unfolds through engaging/embracing our experiencing" (p. xviii). For too long the idea of "knowledge as a commodity or toolbox"—a thing or a substance separate from being—has persisted in mainstream education circles (p. xxiv).

Perhaps our generation is experiencing a *cultural turn*, providing fertile ground in which to seed curriculum and pedagogy that will "expose students to authentic forms of learning that reflect...embodied, dynamic, collective and ecological webs of knowing" (p. xxiv). From ancient Greece to the present, Abram (1996) reminds us, Western philosophy has taught that "human beings alone are possessed of an incorporeal intellect [that]...sets us radically apart from, or above, all other forms of life" (p. 47). Contrary to justifications that human "specialness" gives our species permission for unrestrained dominion over all nonhuman forms, "the human intellect [is] rooted in, and...borne by, our forgotten contact with the multiple nonhuman shapes that surround us" (p. 49). The reasoning mind is one of the wondrous and vital capacities of being that makes individuals "whole," and enriches human lives along pathways of exploration, creativity, and learning. But I question whether the rational model has been elevated to such a degree that educators overlook many *other* resources from which to draw in educating children. I fear that through an over-reliance on rationality for the 'business of schooling,' certain educators and policymakers have dismissed as unimportant much that could be of value for human learning.

We have discussed how objectivity is considered foundational to modern scientific method and is influential in shaping the ways many Westerners come to know and view the world. "Methodization," which separates "knowledge from the somatic...[i.e.,] the emotional, the passionate, the feeling" (Doll, 1997, p. 3) has come at the expense of many other important means of human perception, dismissed as less worthy than those associated with the "rational" center. Separation of mind from *body*, mind from *emotion*, and mind from *spirit* has been transferred from a broad cultural use into the specific ways that we regard education and curriculum.

The "methodization" of the educating process has privileged rationalism as a central discourse driving the bureaucratization of institutions of school-ing. With Molly Quinn (2001), I question if "the notion of curriculum as technique—a functionalist, instrumentalist conception of education" could be transformed into one which "works...in the service of life and the...creation of life worth living" (p. 151). While educating the rational mind is vital, school professionals should not neglect "the aesthetic, existential, creative, imaginative, playful, spirit(ful) aspects" of education for it is through opening the door onto the rich infinitude within human beings that we "begin to shape what we mean by being human" (Doll,1997, pp. 11–12).

It is my opinion that to be a truly generative, creative process, education must recognize the interconnection between all areas of human capacity: not only those born of the rational mind but also those emanating from the energetic drive of the physical body, the seat of mystical intuition, the source of heartfelt emotions, the shimmering seeds of a personal-communal spirit. Why do we limit the educating process to only a *portion* of the multi-dimensional nature of human potential? John Dewey (1934) has said that "in life that is truly life, everything overlaps and merges" (p. 18). For Dewey,

> all the elements of our being...are merged in esthetic [sic] experience. . . in the immediate wholeness of the experience...[which] does not present itself in consciousness as a distinct element. (p. 274)

To dwell only in the house of rationality discounts the rich abundance of alternative ways of perceiving, interpreting, and understanding reality which living affords. For example, Abram (1996) points to the beauty of imagination as being a "way the senses themselves have of throwing themselves beyond what is immediately given" (p. 58). This "beyond" integrates back into the actual through experience for Dewey. He describes the imagination as the commingling of all of the elements of our selves and our experience "at the point where the mind comes in contact with the world. When old and familiar...are made new in experience, there is imagination" (pp. 271–272). Imagination, according to Maxine Greene (1995), serves

> to *awaken,* to disclose the ordinarily unseen, unheard, and unexpected…It is imagination that draws us on, that enables us to make new connections among parts of our experience, that suggests the contingency of the reality we are envisaging. (pp. 28–30; emphasis added)

In cultivating the powers of imagination throughout the educating process, we are educating children toward an ever-awakening sense of their own integrative experiential awareness.

The living of life, in my view, is each being's journey along pathways toward personal "awakenings." This is not to suggest there is a static, actualized state buried deep within and somehow "tap-able" if only we can enter the "right" inner door, but rather, our being is in a constant state of *becoming.* It is never finished. There are qualities of awareness that can always be fine-tuned and further opened onto broader vistas and clearer understandings of who we are. It is these awakenings that increase human capacities for negotiating life's passages. Our being *is* the experience of a constant process of creation. I see this potential for creation at the core of what educating children could be about, as action framed within creative process (Dewey, 1934). In Dewey's terms, "the ideal…emerges when the imagination idealizes existence by laying hold of the possibilities offered to thought and action" (p. 33). Envisioning "the possible" within "the actual" (Garrison, 1998, p. 31) is imagination's promise to education, which acknowledges the constant generativity of further understandings, of negotiating passages across chasms of contrast and difference, bridging incommensurable vocabularies in search of new languages, new ways of communicating, and the means toward further dialogue.

With bell hooks (1994), I wish to consider education as an "art form" that explores the possibilities of opening up the discussion on what it *could* mean to school, what it *could* mean to educate. I would suggest that education is more than the production of "accomplished test-takers" or "efficient and docile employees for business" (Pinar et al., 1995, p. 848).

It is also about helping children to flourish as human beings in all of their fullness,

> think[ing] and act[ing] with intelligence, sensitivity, and courage in both the
> public sphere—as citizens establishing a democratic society—and in the private
> sphere, as individuals committed to other individuals. (p. 848)[1]

And it would move away from old patterns underlying Western worldviews such as patriarchal and dualistic thinking, which have limited our ability to widen our vision of schooling, one which would encourage a sense of children's organic connection within the complex body of all things constituting the earth.

I am interested in the idea of ecological interrelationality as a viable heuristic through which to begin re-"thinking" the separation that under-girds the "methodization" of schooling. I am particularly drawn to ecofeminists (Kaza, 1993; Macy, 1991a; Merchant, 1996; Ruether, 1996; Spretnak, 1997; Warren, 1993) whose work explores the sacred nature of the human interrelationship within the ecological world, a relationship fused throughout—and which also fuses—every planetary form within a living process. I also resonate with strands of postmodern thought (Bohm, 1985; Deleuze, 1977/1987; Doll, 1993) as a means of disrupting modernist discourses which separate human beings from the full experience of their kinship with other people and with the ecological world. Therefore, I will explore the potentialities of an ecological vision of schooling (Bowers, 1993; Fox, 1995; Macy, 1991a; Miller, 2000; Smith & Williams, 1999; Warren, 1996), both looking at its problems and also its possibilities for re-envisioning curriculum and education as an *ecospiritual praxis* (Spretnak, 1997; Warren, 1993). An ecospiritual praxis for understanding curriculum would focus on "ecological *processes,* such as interdependence, sustainability, partnership, flexibility, diversity, and co-evolution" (p. 122). This type of approach is sometimes referred to as "ecoliteracy," but the connection made to the term ecology "should not be...confused with far more narrowly focused 'environmental education'" (Publications from the Center for Ecoliteracy; as cited in Spretnak, 1997, p. 122). I do not mean "ecological" in the sense of environmental management—typically human-centered and often ascribing only instrumental *use-value* to nature (Spretnak, 1997)—but rather, as a worldview that recognizes the natural world as having *intrinsic* value,[2] defined as the "value that a being or thing has in and of itself, independent of human valuation" (Plumwood, 1997, p. 371). In terming it ecological, the vision for curriculum that I am proposing recognizes the mutually sustaining character of the human-earth relation and respects the inherent value of each part within that process. It posits that an ethic of earthcare[3] could be fundamental to a wide-scale consciousness-raising through children in school and could help to heal the "split" that humans have inherited, at least since Descartes. Abram (1996) reminds me that

> every theoretical and scientific practice grows out of and remains supported by the forgotten ground of our directly felt and lived experience, and has value and meaning only in reference to this primordial and open realm. (p. 43)

With Charlene Spretnak (1997), I wish to push for an ecological postmodernism that would re-member the severed realms of who I am, my heart, my lungs, my body. I wish to open myself to my own intuitive regions, to my emotions, to my sense of spiritual connection, to my own imaginative wanderings, which bring me to places for feasting on vivid colors, sounds, impressions—the experiences of my own sensate receivers. Because of the passion of his prose, I am compelled to offer this rather lengthy quote from David Abram's (1996) *The Spell of the Sensuous*, through which he describes the vitality of humans' connection with their own interior worlds and with the corporeal world around them. He writes:

> Humans are tuned for relationship. The eyes, the skin, the tongue, ears, and nostrils—all are gates where our body receives the nourishment of otherness… For the largest part of our species' existence, humans have negotiated relationships with every aspect of the sensuous surroundings, exchanging possibilities with every flapping form,…textured surface…shivering entity…All could speak, articulating in gesture and whistle and sigh a shifting web of meanings that we felt on our skin or inhaled through our nostrils or focused with our listening ears, and to which we replied—whether with sounds, or through movements, or minute shifts of mood…And from all of these relationships our collective sensibilities were nourished. (p. ix)

I wish to nourish my own sensibilities and those of the students who enter my classroom. I wish to offer us all a forum where we may come to know one another and the world around us. I wish to incorporate a sense of the sacred within those relationships, the shimmering web of connection that infuses us all with/in the life force of creation.

In keeping with this commitment, my methodology in this work entails a weaving: the laying down of threads, parallel lines, layers of weft and warp, circle and dimension. It is an analytical project as I simultaneously critique the model of discourse within which I write. By intertwining theory and personal voice, I hope to appeal not only to "reason" but also to the aesthetic perceptions felt at the corporeal, sensing level of the body; within the emotive, intuiting heart; and in the visceral knowing way of the human spirit. In order to bring theory to life, I will bring "life" into theory. While modernist tendencies have been to privilege ways of understanding that fall into categories sanctioned as verifiable through "objective" observation, feminist theory has shown that ways of knowing are shaped "by the assumptions, values and interests of a culture" (Longino, 1989, p. 212; see also, Harding, 1991; Jaggar, 1989; Keller, 1985) within a social and historical framework of time and place. Many feminists and postmodern theorists would suggest that the notion of utter objectivity is a "myth" we cannot afford. A double-bind that has challenged many feminists is that to write "critiques of reason from the margins, from the place of exclusion" is also to participate in the very philosophic discourse that has marginalized "women...in the first place" (Harvey & Okruhlik, 1992, pp. 11–12). Recourse lies "not in the rejection of 'rational discourse' as an irredeemably masculine construct but in increased participation by feminists in the making of art and science and the discourses that interpret them" (p. 18).

Therefore, I will intentionally interrupt the linear stream of the more formal "objective" language valued within academic tradition, and move in and out of personal narrative through the body of the text. In using personal voice, I must make some qualifications at the outset. As a genre within contemporary discourse, autobiography has been called into question by feminists and postmodern theorists (Bergland, 1994; Butterfield, 1994; Gilmore, 1994; Miller, 1997; Munro, 1996 and 1998; Pinar,

1985 and 1998) for ways it has traditionally privileged, both, the Western, white-male voice and extolled the idea of the "autonomous individual." Historically, the majority of autobiographical writing has reflected the white, Western, heterosexual, male perspective as "poet, scholar, citizen, politician, and hero" (Gilmore, 1994, p. 1). This casting of the "representative man" (p. 4) has excluded many "Others" as being less important. Such writers as Augustine, Rousseau, and Thoreau are representative of what has been valued within traditional studies of autobiography and autobiographical critique. These *particular views* have been considered as legitimate autobiographical form, while *others have been excluded*. This tendency for those working within the genre to valorize the autonomous (male) "individual" at the expense of Others, "denies the lives of millions…[and] masks the ways in which we are constituted by language and positioned differently there—depending on race, class, gender, or ethnicity" (Bergland, 1994, p. 161). This marginalization is further problematic because the anthropocentrism and androcentrism underlying many of the world's ecological crises is fed by the cultural beliefs and practices based in large part on the Enlightenment ideal of individualism (Bowers, 1995).

Feminists, poststructuralists, and a variety of Others are disrupting canonical "truth" claims by writing critically from the margins in order to confront traditional contexts in which autobiography has been normalized. Bergland (1994) offers a means through which autobiography may be reconsidered: for its political and ideological uses, as naming an autobiographical self that "must be understood as socially and historically constructed and multiply positioned in complex worlds and discourses," and by utilizing and exploring "alternative strategies for reading and understanding autobiographies" (pp. 131–133). Autobiography is being reclaimed as a site for disruption by women, people of color, gay men, and lesbians who, in speaking their alterity, contribute to the re-conception of autobiographical theory and notions of "identity." They are replacing the idea of a "unitary individual" with multiple subjectivities and providing alternative views on a "relational self" (rather than "essential self"), which is characteristic of postmodernism's influences.

Autobiographical writing allows me to speak in and through many languages, many voices, move through worlds of endless possibilities. It provides a space to be filled by the "as yet," the "unknown," the recounting of tales, of musings and murmurings, textual configurations that can have the "effects of transforming life...into a text" (Olney, 1980, p. 6). Autobiography can speak in the language of dreams—weavings and wordings captured in the night out of fruitful slumber onto reams of white paper. Pinar and Grumet (1976) have said, "We must lay in waiting for ourselves. Throughout our lives. Abandoning the pretense that we know" (p. viii). This pretense to knowing has driven our world with its claims to authenticity, viability, reliability, categorization, qualification, legitimation, and mechanization. Education would be heir to this epistemological inheritance. Curriculum has been defined as the *knowledge worthy of passing on to the next generation.* There is no place within mainstream curriculum "knowing" for the language of dreams because, as Mary Doll (1995) attests, such language is "too ephemeral, lacking text in the linear-sequential sense of text" and also because it is "dangerous, too personal" (p. 68).

Our thinking portends, in academe, toward the rational-empirical frame of Western scientific thought with its roots in Enlightenment precepts. The academy undervalues the stuff of dreams—imagination, intuition, emotion. The academy values reason. There is an old argument which goes: "values and emotions may be important in the context of discovery but are insignificant in the context of justification" (Harvey & Okruhlik, 1992, p. 8). In the name of justification, rationality has been separated from and privileged as "above" the wider gamut of many other human traits and capacities, including, for example, emotions, values, imagination, dreams, and intuition. Yet, these are "aspects of human knowing inseparable from the other aspects" (Pinar & Grumet, 1976, p. 137). Western seventeenth-century science left us defining "values and emotions" as "variable, idiosyncratic, and subjective," whereas "trustworthy knowledge could be established only by methods that neutralized...values" (Harvey & Okruhlik, 1992, p. 10). Such an attitude is likely to consider knowledge as a determined and static property to be revealed through the "trustworthy" means of "right reason" and experimental method. The screening out of values and subjectivity in order to "guarantee theoretical objectivity" is grounds for the language of science (p. 68).[4]

A difficulty in using autobiographical theory arises with the use of universals, mentioned earlier, when the author generalizes his or her own perspective as if that one view could apply to an entire category of people. I will avoid speaking from categories wherever possible to avoid collapsing a wide range of unique and particular positions, losing the richness of their differences. With that caveat, however, there are times when speaking in categories is necessary to communicate certain ideas or to defend a position. When Eric Fromm (1955) asks, "can a *society* be sick?" (p.12) or when David Abram (1996) questions how *we* have "become so deaf and so blind to the vital existence of other species, and to the animate landscapes they inhabit" (p. 28; my emphasis), these are queries as to the state of a larger condition, set against an implicit hope for a better world. The constructs "society" or "we" are necessary in asking these sorts of questions.

The autobiographical interludes woven through my work allow me to write in the language of personal voice, a language based in "lived experience," experience that excludes *neither* rational thought *nor* the more aesthetic possibilities for coming to know the world. We are creatures of *both* reason *and* emotion, mind *and* body, matter *and* spirit. The language of personal voice lends a dimension to academic writing that cannot be filled by expository scholarship alone. The quest for "truth" has led people astray, "lost from ourselves" and has resulted in an estranging alienation (Pinar & Grumet, 1976, p. 3). The scientific "conception of objectivity...has allowed scientists [and many mainstream educational researchers] to deny their own biases and to ignore the centrality of lived experience" (Harvey & Okruhlik, 1992, p. 71). Personal narrative admits to biases upfront as parameters within which the "truth" of a circumstance comes to be constructed with the context of time, place, and circumstance.

The stories I tell draw heavily on my sense of kinship with the natural world, which is for me a spiritual connection. I do not wish by that to valorize or exalt nature as separate or above culture, re-inscribing a well-worn split that I see as an illusion. Instead, I wish by my storytelling to convey my personal interpretation of the human/nature connection as being a part-to-whole relationship (which is certainly not my view alone) and to offer my own sense of *living* that relation. Personal stories can fill in gaps toward more thorough and well-rounded representations of "factual" information. Personal narrative acknowledges the importance of humans' embeddedness within particular contexts of time and place. Autobiographical writing recognizes that people are situated within specific geographic settings, sharing cultural traditions, recounting oral histories, the stories that make up our lives. All in all, I want to pull and draw and recombine the threads of meaning into some larger form, rife with vivid details, colors, textures, multiples of interpretation and possibility.

And so, I "construct a line of flight" upon and through a *medium* called language, a habitat wherein humans arise and have their being. A "medium" is a place of *betweens*, from reference-point-to-reference-point, cracks in the pavement through which new shoots may arise (Deleuze, 1977/1987), places for birthing "all the combinations which inhabit us" (p. 4). The word *medium* has broad implications for the ways that humans live within language, being defined as "a surrounding environment in which something functions and thrives;…a specific kind of artistic technique or means of expression;…occurring or being between two degrees, amounts, or quantities;…intermediate" (*American Heritage Dictionary*, 1996, p. 1121).

Alongside the signifiers, the reference points, language is a place of *betweens*. As humans, we live in languages—multiple languages—which are more than words, but also, thoughts, ideas, sudden awarenesses, bodily postures and gestures, nods and nuances, mental images, tugs at the heart. In the entire enveloping world of sensual perceptions (Abram,1996), alongside a broad spectrum of ways by which humans come into consciousness and cognize the world, all is signified within the embrace of language. Language is a landscape of becomings—not to be, the IS, but becom-ing used as an infinitive, as on-going creation, dissolution, equilibration of life processes—*limitless* becomings. Deleuze 1977/1987) has said that

[b]ecomings belong to geography, they are orientations, directions, entries and exits.... To become is never to imitate, nor to 'do like,' nor to conform to a model.... There is no terminus from which you set out, none which you arrive at or ought to arrive at.... Things never pass where you think or along the paths you think. (pp. 2–4)

Language—life—is that way. If language is life, there is no such thing as a perfect sentence that we come to at last, no exact word, "only inexact words to designate something" (p. 3), only nodal points between the betweens, which are not really points, but pointings-to, and which can never say, exactly, 'what is' or 'how things really are.' And so all we can really do is to construct a line of flight, a not-knowing, because "things never pass where you think, nor along the paths you think" (p. 4). In attempting to think "an active pluralism" (p. xii) Deleuze uses the metaphor of the rhizome, "a multiplicity of interconnected shoots going off in all directions" (p. xi). He contrasts this rhizomatic image with the metaphor of the tree, long a symbol within the history of Western philosophy denoting a "hierarchy of root, trunk and branch" (p. xi). For Deleuze, the Western philosophical idea of "rationalism" has pointed always toward first principles: "the One, the Whole, the Subject" and one then goes about the task of making life "conform to their requirements" (p. vii). The most difficult aspect of this stance arises when "one sees rational unity or totality turning into their opposites, or the subject generating monstrosities" (p. vii). Instead of the Whole or the Subject, Deleuze shifts the focus to one of *multiplicities*, "designat[ing] a set of lines or dimensions which are irreducible to one another" (p. vii). Within a multiplicity, he says, it is not the reference points i.e., the "terms or the elements" that count, but what lies in "the between, a set of relations that are not separable from each other. Every multiplicity grows from the middle, like the blade of grass or the rhizome" (pp. vii–viii).

And so I speak in multiple voices, telling stories of multiple worlds, often traveling within incommensurable vocabularies, a becoming-multilingual. Becomings are "assemblages," according to Deleuze (p. 51), always collective, which bring into play within us and outside us populations, multiplicities…Structures are linked to conditions of homogeneity, but assemblages are not. The assemblage is co-functioning…being in the middle, on the line of encounter between. (pp. 51–52)

The conjunction *and* is a word of connection that allows for bridging impasses of either/or, where "each encounters the other a single becoming which is not common to the two…but which is between the two…outside the two, and which flows in another direction" (p. 7). The conjunction *and* sends "dualism off course" (p. 57): black-and-white-becoming, male-and-female-becoming, modern-and-postmodern-becoming, each a "single becoming, a single bloc, an evolution…which flows in another direction" (pp. 3–7). From one across to another who is different, those happenings in the *between* give rise to the birthing of something new altogether—creation. It is why Deleuze says "the question 'What are you becoming?' is particularly stupid. For as someone becomes, what he is becoming changes as much as he does himself [sic]" (p. 2). Becomings are the things which are "most imperceptible" (p. 3), becomings are about *not knowing*" (p. 5). Yet, living in language, we try and capture these becomings in words when we try and delineate what and how things are, make them concrete, make them solid form.

And so it is coming clear with this work that I am creating a line of flight, a journey 'in' to the ground of not-knowing. The hour is late (or could it be that it is too early) and I engage in the process of producing a document, an undertaking that is sometimes painful, as it tends to relegate creativity onto a template of expected outcomes. I re-visit this drawing board again and again, fresh from what seemed a pathway to follow, only then to transform into a false start, a tangent, from which I must once more return. And so here I sit again, at my desk before the window, arranging scraps of notes and the coffee cup, just so, calming my hands from the quiver inside I feel from the fear of "not-knowing." Yet, what is to fear? Isn't this "not-knowing" the very alternative that I propose for curriculum—a "loosening up" of the language of determinacy and absolutes, so long the currency of exchange in institutionalized discourse? It is a question of *how* to say what must be said, when *the saying of it* risks the very totalization that I seek to disrupt, and when it is about so much more than words can come near. Emerson's dilemma (1965) is my own:

> I, cold because I am hot—cold at the surface only as a sort of guard and compensation for the fluid tenderness of the core—have much more experience than I have written there, more than I will, more than I can write. In silence we must wrap much of our life, because it is too fine for speech, because also we cannot explain it to others, and because somewhat we cannot yet understand. (p. 190)

Still, we move on, and so we reach out to communicate. But, tell me how does one *speak* the freshness of the wind? How to name and categorize and still expect to come away invigorated and re-newed? Where are the words for that *experience* of exhilaration? How to write in the language of the fluid, the indeterminable—when to name, to label, is to pin down and make static, indeed, to create a form? And even still, isn't there need for *some* structure, *some* form, when carving out pathways for young ones to follow? Isn't there need for some balance *between* freedom and stability, a way to stand on firm ground, while still resisting the desire to unify, to totalize, to resolve? Doesn't one always, ultimately, return to this ground of "not-knowing?" It seems that it is here, in the between-spaces, where my journey must begin, and rightly so. For that is my project, and also my dilemma, to speak for what can never be spoken yet can be approached through the dynamic possibilities that open space affords, ever-new capacities for emergence and change. Am I not intent upon moving toward the mystery, the uncertainty, the richness of *hope* followed like a path through the unknown?

Perhaps the metaphor of "carving out pathways" is a good one. As a child in Mississippi, I would walk long days in the woods near where I lived. Leaving the neat rows of homes and green suburban squares lined with concrete and telephone wire, I'd enter into the tangle of underbrush, always coming onto a clearing that I'd made. And as days would pass, I'd create winding pathways in through the brush, sometimes following animal runs, while at others breaking off in new directions, where a turn might be taken around an old stump, or an abrupt left was followed along the back of a fallen pine. Perhaps teaching can be just such an adventure of discovery. And the structure comes, perhaps, in a return onto those clearings, in the consistency of recursion (Doll, 1993), by coming back again to that base, the foundation on which we build. And perhaps one could say that structure comes in *partial ways* as do the unexpected turns which rise out of particular contexts; abrupt reconnoiterings that are necessary when there's a deep chasm to cross or a manuscript to pour-out-of-your-heart in a few months' time.

And so I own up to uncertainty. There are no exact sciences. Even in the best of scientific circumstances, there is always the potential for some wind to blow up and change the whole approach; some chink in the observer's armor that will cause the colors to run, to bleed through onto a different page; some glitch that will change the language game (Brill, 1995, pp. 13–17) and thus the rules we thought were firm. And so, the best we can do is to construct a line of flight, never certain where it leads, because "things never pass where you think or along the paths" (Deleuze, 1977/1987, p. 4) you expect. And in expectation much can be lost. Expectation is about the future or the past, which is to separate life's continuum into artificial segments, "clean-cut states side by side" (Bergson, 1911) rather then "a flux of fleeting shades merging into each other...a continuity which unfolds" (pp. 3–4) always within the now. Expectation is about control because it places thinking in the future, or else, somewhere in the past—either attached to some outcome or resisting it like crazy—and the danger, there, is in not experiencing the present, the Now moment. Becomings are always and only in the present moment. The present moment is about geography, situation, context, place. It recognizes the sensual world of the body, drawing also on imagination, emotion, and the intuition that we may be spiritual beings, embedded somehow within a larger network of associations, a micro-macrocosmic view, positionings in places of relationality: connections and betweens. When all is said and done, the landscape of our lives is made of connections and betweens. Nodal points connect the betweens in a continuum of relations, never an ending, or a beginning that is clean cut, no single origin nor terminus—not that we humans can know.

And so it is the between that is rich and fertile, a prime medium for creation. It is the *between* that is interesting because it is neither *a* nor *b* but something new altogether arising in the middle, "a single becoming which is not common to the two, which has its own direction, a bloc of becoming, an a-parallel evolution" (Deleuze, 1977/1987, p. 7). And thus it is in the between-spaces that I sow these seeds of theory—in the margins where the center gives way to open ground for new growth to flourish. And I begin with a beginning, written, yet, before the beginning you've just read, speaking in and between many languages: speaking in a voice grounded in a history that precedes this beginning, a language of my own history, my own memory, my own sense of "duration" as "that prolongation of the past into the present" (Bergson, 1911, p. 17), "...rolling upon itself as a snowball on the snow...chang[ing] without ceasing" (p. 2). Across these beginnings there are languages grounded in Western philosophy, feminist epistemology, ecological postmodernism—at times—"non-homogenous sets" (Deleuze, 1977/1987) taking off in lines of flight out of the variations among and between their differences, other ways of knowing, incommensurable vocabularies (Lyotard, 1979). I am reminded that the "difficult part is making all the elements of a non-homogenous set converge, making them function together" (Deleuze, 1977/1987, p. 52) and understand that the intent is not to homogenize but to "speak *with*, write *with*" (p. 52), because in doing so, I do not need to mistake myself for the Eskimo boy going by, but, perhaps, momentarily putting myself "in his shoes, [I]...have something to assemble with him" (p. 53), something between us both.

And so I speak in multiple languages, anchored both, within "the betweens," the clearings, and also to the reference points—realizing that there are times when some exchange is made, or when a shift in ground occurs, an exchange of positions, where the *between* is signified and the center takes the margins. And all the while, I seek to gain my footing and to find equilibration, some sense of positionality, of place, from which to then break off in new directions, finding structure in partial ways and giving voice to new languages, new colors, new patterns alongside those which were there before.

I was a weaver early on and here weave for you a vision, an assemblage[5] of stories, both personal and theoretic, intent upon constructing a line of flight, a "becoming" that is neither one nor the other, an a-parallel evolution. From "between" the personal and the theoretical, I will consider how a relationally based, ecological spirituality and curriculum theory might intersect, dance, collide, assemble, co-mingle, unfold within a "telling." The stories I am interested in are those which sow seeds of life, understanding, creative initiative into hearts and minds and souls of educators, and also, of young ones venturing onto their own life paths with "fresh minds ."[6] As educators, it is ours to work with young ones, to experience with them, grow with them, be with them in relationship wherein the living happens on both sides of the desk, an assemblage, a teacher-student becoming, an a-parallel evolution. My desire is for writing and thinking and living theory. Especially *living* theory—because it is in the living that theory comes to be—where it moves beyond the abstract, no longer symbols on a page, a map of reference points by which we may guide our lives in schools. And it is in bringing theory to life that we may breathe life into theory. It is in living theory that we may focus figure into ground and ground into figure, no longer overly-determining any point as the center but an open field wherein tree and grass and rock and sky may exist as what and even who they are: individual expressions within some larger ecological system that manages, somehow, to find equilibration among the order and disorder of its varied regions. It is in *living* theory that we may foreground the between-spaces, *between* the reference points from A to B, from here to there—rich and fertile soil wherein the dynamic interplay of life realizes its most creative potential. In the movement "between" there is less chance for the center to solidify into a totalizing structure because it is an interval that encourages dynamic play, an open ground as fertile as the earth itself, generative with ever-abundant seeds for fresh starts and new beginnings.

What I propose, then, is a re-cognition of humans as sensing, reasoning, expressing, creating beings. Each is a particular and viable particle within an ecological matrix of mutually sustaining relations and each nexus of relations is vital in constituting the whole in its continuing evolution.

Notes

1. According to Pinar et. al (1995) "feminist theory has shown [that] the two spheres are distinguishable in concept only" (p. 848).

2. I incorporate Plumwood's (1997) use of the term "nature" which includes all "nonhuman animals, ecosystems (urban and otherwise), and nonsentient natural things" (p. 370).

3. Hallen's notion of "earthcare" (1987; as cited in Merchant, 1996, p. 206), as well as other examples of ecocentric perspectives are discussed in chapter 4.

4. The rationalization processes of scientism and reasonism are discussed in chapter 2.

5. Deleuze (1977/1987) defines an "assemblage" as a "multiplicity which is made up of many heterogeneous terms and which establishes liaisons, relations between them, across...different natures. Thus the assemblage's only unity is that of co-functioning: it is...symbiosis...not successions, lines of descent, but contagions, epidemics, the wind....the set of the affects which are transformed and circulate in an assemblage of symbiosis [is] defined by the co-functioning of its heterogeneous parts" (pp. 69–70).

6. Shunryu Suzuki (1970/1998) describes an ideal mind within Zen practice as one that is "soft and open enough to understand things as they are" (p.115). He terms this "beginner's mind" (p. 21) and says that "[i]n the beginner's mind there are many possibilities, but in the expert's there are few" (p. 21).

Part Two

All phenomena are mind…Mind is matter, matter is mind. Matter does not exist outside of mind. Mind does not exist outside of matter. Each is the other. This is called the "nonduality of mind and matter."

—Zen Master Dogen,1200–1253

Toward a Reweaving

Humans' "Othering" of the natural world is a deeply ingrained practice, "rationalized" as it has been through the philosophical mindset of patriarchy. The human species and all nonhuman forms on the planet, have co-evolved over a 14-billion-year process. Yet many people operate as if planetary life existed to support and sustain human well-being alone. Separation, dualism, and power-over ways of knowing fuel a human-centered attitude that exalts our own species above the rest of the natural world. With Patsy Hallen (1987), I feel that what is needed is an "ethic of earthcare" (as cited in Merchant, 1996, p. 206), which acknowledges the reciprocal nature of the human/earth relation. Rather than a hierarchical perspective, it stresses eco-centric balance between the human species and the ecological world. If an ethic is an "approach to living" (Palmer, 1983, p. 51), I have come to an approach to living by means of "an approach to knowing: epistemology" (p. 51). Part 1 explored relationality as an epistemological question.

I now widen the discussion of *how we know* to include *how we live*. Ethical questions are questions grounded in place, history, culture. They move a discussion of knowing into the realms of agency, action, doing. Ultimately, how we know and how we act cannot be divorced, for each stands balanced within the dynamic play of mutual interaction, i.e., "mind and world...[are] dependently co-arising" (Macy, 1991a, p. 66). And it is from this interactualization of knower and known, this dependent co-arising *between* subject and object that something emerges, an a-parallel evolution that is not one nor the other but something new altogether (Deleuze 1977/1987). Difference is the fertile ground from which new worlds arise. And so I move into an exploration of relational ways of knowing by way of an inquiry into difference.

Chapter Three

All becoming has needed me. My looking ripens things and they come toward me to meet and be met.

—Rainer Maria Rilke, 1899

Ways of Knowing "Others": Living the Border Life

Four miles south of Louisiana State University on Highland Road is Mt. Hope Plantation, with its fading opulence offered to garden weddings and ladies' luncheon clubs on springtime afternoons. My small house is just across the street on the western side, a stone's throw from the plantation's aging white gate. Here are two cottages side by side; I live in one and the other has stood empty for years. They are 15 x 20 wood-frame structures, moved to this site as housing for workers in the 1940s, from up the river near the Old State Capitol at Baton Rouge. The structures remain covered in a broken, mud-colored siding, cloaked with light-green algae over walls and deck, a soft, verdant blush that plumps-up like a layer of thin velvet on rainy afternoons. From the road, there is a gravel turn-in and then the land begins to drop down a steep, grassy incline to Bayou Fountain sixty yards below. At that point ground gives way to swampy, lowland-wood marsh and extends a few miles west to the Mississippi River. From the road above what's visible is a crude tin roof and the top of a pair of fading white shutters. The cabins are perched on this hillside, flush with the soil on the front, but suspended in the rear 4 or 5 feet above the sloping ground as the land drops toward the bayou.

My cabin sits in a clearing, while the other one can scarcely be seen from any vantage, immersed as it is within a thicket of overgrown brush, trees, and a large stand of emerald green bamboo, that has gradually thickened year by year. The 30-foot distance between the cabins marks a narrow bird-run from the Mt. Hope grounds across the road, down the incline to the bayou and the thick, swampy woodland beyond. Every manner of Louisiana fowl can be seen or heard as they move in and out of these backwater woods. We cohabit this "border world," the Highland Road, verging the ground at the edge of a metropolitan city of 240,000 people and a teeming hinterland of water oak, palmetto, and meandering inlets.

Ecologically speaking, there are regional zones in the natural world—sometimes given names or labels—characterized by distinct physical conditions, and populated by communities of certain kinds of organisms. Florence Krall (1994) refers to the area *between* such distinctly defined regions as the "ecotone," that is, a "boundary between two natural communities where elements of both as well as transitional species intermingle in heightened richness" (pp. 3–6). An ecotone is a border world with a rich complexity that comes, in part, from a commingling among a variety of diverse forms cohabiting a common area. It is a margin zone, a "crack in the pavement" through which something new altogether may arise, offering new possibilities for creation, forms of life, ways of thinking or knowing the world, a-parallel evolutions (Deleuze, 1977/1987). Such margins are often chosen by animals as places for raising their young, since they are habitats "where the greatest variety of cover and food can be found" (Krall, 1994, p. 4). The "edge effect" of an ecotone suggests to ecologists "the complex interplay of life forces where plant communities, and the creatures they support, intermingle in mosaics or change abruptly" (p. 4).

Thoreau writes of living a "border" life between society and the natural world (Walking, 1914) and describes border zones as biologically diverse areas between civilization and wilderness that are often the "richest in spiritual and intellectual possibilities" (as cited in Payne, 1996, p. 42). As a social/cultural metaphor, ecotones can be boundaries that separate and divide, as in a front line battleground. Conversely, they may "be dwelling places that serve to connect rather than separate,…rich and dynamic transitional zones...[that] provide great learning as well as suffering" (Krall, 1994, p. 4).

Border worlds are *between* regions, marked by difference. They are often undetermined spaces, prime locations for resistances to emerge; new strains of life, dissident voices speaking their particularity. Difference is a way of knowing that arises from between the 'knowns,' the 'certainties,' the reference points of A or B. Difference is a way of knowing that is born of "not-knowing," emanating from terrain that has not been overly determined as the center or "the truth." As a way of knowing, difference connotes a willingness to let go and let be. Not overly determining any one point as truth leaves a space for all the possibilities that openness affords. The space of difference is a fruitful realm from which new worlds arise. In order that it bear fruit, difference need be acknowledged and accepted with a tolerance for the "Other," who is not self-same.

Ecofeminist theory arises from just such a border world. Determinate viewpoints staked out as "truth" can limit a worldview within parameters of either/or—culture/nature, male/female, self/other. Much of modernist thought has embraced positions of certainty, pushing to the margins those people or ideas not of the dominant group. Women, people of color, Jews, gay people, and the poor are some of the people who have been Othered— marginalized. Ecofeminism draws together a complex variety of those named as Other, giving "voice to a subversive politics, aware of its own situatedness and transitionality" (Salleh, 1997, p. ix). Ariel Salleh (1997) situates ecofeminism as a "'womanist' rather than a feminist politics...theorizing an intuitive historical choice of re/sisters around the world" (p. ix). She includes within its "groundswell men and women who would not necessarily name themselves 'ecofeminist' but who act in ways that promote the same complex of objectives" (p. x). Salleh says that ecofeminism

> transcends differences of class, age, and ethnicity between women...and reaches for an earth democracy, across cultures and species...refram[ing] environment and peace, gender, socialist, and postcolonial concerns beyond the single-issue approach fostered by [the] bourgeois right and its institutions. (p. x)

Ecofeminism, like feminism from which it derives, is non-unitary and at times manifests in contradictory perspectives. There are multiple versions and particularized positions, lending richness and complexity—and also divergent tensions—to ecofeminist cultural, political, and spiritual theory. Resisting univocality is seen as a strength by many ecofeminists (Quinby, 1990) who recognize and embrace the coexistence of conflicting viewpoints rather than seek to flatten differences into a mono-vocal story. There are women of the African spiritual tradition who bring their cultural ways to the public by attempting to integrate "the political and the spiritual with contemporary feminist and Black power politics" (p. 112). There are Native American women organizing politically, while they are trying to make their traditions known, hoping to counter the exploitation of "their lands...by developers...and industry" (p. 112). There are multiple goals of diverse and particular women working within their own struggles, their own resistances.

Even in its plurality there are some common characteristics across different areas of ecofeminist theory. One shared commitment is "to the elimination of sexism, i.e., the power and privilege of men over women...wherever and whenever it occurs, and to creating practices and theories which are not male-gender powered and privileged" (Warren, 1993, p. 122). Ecofeminisms interrogate "features of a patriarchal conceptual framework" such as hierarchical and dualistic thinking, "power-over conceptions of power," and "a logic of domination" (p. 123). Ecofeminism and other feminist theories have made important contributions to cultural thought. Beyond exposing gender-based oppressions, feminist discourses have also critiqued patriarchal domination in the forms of racism, classism, individualism, and naturism,[1] among others. The "power-over"[2] frame of patriarchy does not limit its reach within the bounds of gender. And for this reason, feminist critique is a necessary component of an ecological perspective seeking to end dualistic, hierarchical thinking that leads to environmental degradation. In as much as ecology explores the viability of relationships, ecofeminist theory provides the language to disrupt discourses of domination within many realms, pointing to ways that oppressions are predicated on a basic premise: separation.

Domination and hierarchal control have often been justified in the name of what is "natural" or "the order of things." For example the widespread domination of women as "naturally" being the weaker sex; discrimination against gay men and lesbians in accordance with what is "normal and natural" (Ramazanoglu, 1993, p. 33; see also Pinar, 1998); the perception of blacks or ethnic groups with a skin color other than white as being "inherently" inferior—these are among countless examples of people having been systematically excluded or devalued according to the rationale of "natural law." According to Ariel Salleh (1997), the triad of oppressions of "Man over Man,...Man over Woman, and...Man over Nature" (p. xi) cause her—and many other ecofeminists (Merchant, 1996; Plumwood, 1996; Ruether, 1996)—to question whether, "like a Boromean Knot" they may "only be dismantled together" (p. xi). For Salleh, gender is the "lowest common denominator of all dominations" positing as she does that "environmental struggle is socialist struggle is feminist struggle" (p. xi).

Women and nature have historically been conceived "in terms which feminize nature, naturalize women and position both...as inferior to male-gender identified culture" (Warren, 1993, pp. 122–123). An androcentric separation undergirds the twin dominations of women (sexism) and nature (naturism) as inherent in epistemological structures, in the ways of knowing constructed within patriarchal Western society. Just as the "logic of domination" has Othered women, forms of life within the natural world have been judged as important only insofar as they serve a purpose considered necessary for humans. This anthropocentric regard for life causes many people to devalue biological realms, reducing nature to the status of a mere resource for human consumption. This skewed "logic" is a denial of the vital nature of each form of life, of its value in its own right, and also its contribution to the balance of the larger ecosystem. Narrowly focused, human-centered thinking fails to meet its own self-serving ends by disregarding the human species' dependence on the health and balance of the natural world. To destroy the earth is to destroy all forms of life upon it, including our own.

As a majority of women around the globe live their lives "at the bottom of a hierarchy of oppressions" they inhabit a complex and "contradictory space," a border-world where Woman and Nature come together (Salleh, 1997, p. xi). Catherine Keller (1993) notes that "feminist theory has mapped well the long history of religious, philosophical, and scientific projection of 'nature' onto woman's body and of woman's body on the Earth" (p. 291). The position is contradictory because the conflation of woman with the nonhuman world is at once the distinction that Others and subordinates women. However, it also puts women into prime positions "to resist exploitation and to care for the environmental community upon which their welfare and that of their families depend" (Ruether, 1996, p. 4). According to Ynestra King (1990), some feminists disagree (Simone de Beauvoir, for example), questioning if the "woman/nature connection [is] potentially emancipatory or whether it provides a rationale for the continued subordination of women" or if it constrains women in a "primordial realm…that is bound to reinforce sex-role stereotyping…[and] gender differences" (p. 110). Hallen (1987) would argue that "history and socialization" have placed "women in a strategically important position to develop a desperately needed ethic of eathcare" (as cited in Merchant, 1996, p. 206). She posits that "women can reclaim their historical past without being chained to it and can choose their historical future" (p. 206).

Patriarchal ways of knowing have led to cultural practices which are implicated in the serious environmental devastation going on around the world, a condition environmental advocates are working hard to eliminate. Before turning to some of those struggles, I will discuss the environmental movement within the United States, looking at some of its history, and the contributions environmentalists and deep ecologists are making to theorize those efforts.

The Ecological Landscape: Some Historical Terrain

[T]he predominant American worldview has been, and continues to be, an anthropocentric one, that sees the earth as a virtually limitless storehouse from which humankind can extract resources and where it can dispose of waste and alter the landscape with little concern for the ways in which those actions will affect the local or global environment.

—David G. Payne, 1996

Ecological awareness as a sensitivity to the mutually sustaining nature of human-earth relations, has become more widespread within the last century (Payne, 1996). Early European settlers came to North America with hopes for an abundant new life and also fears of what they construed as a foreboding and dangerous land. Viewing cultivation and development of the terrain as paramount to their survival, they systematically, and often wastefully, destroyed forests and animal life in a sense of urgency to transform what they considered to be a hostile environment into a place fit for human habitation. Puritan settlers equated the taming of a "savage" land with a "sense of spiritual purpose," as fulfillment of God's divine "errand into the wilderness" (p. 1). They viewed the holy nature of their task as vindication for their colonialist appropriations of territory, believing that to allow land

> to "lie waste" through a lack of cultivation and development...[was] an abrogation of their duty to subdue [it] as commanded in Genesis 1:28. This duty, they believed, had been ignored by the Indians [sic], and thus they justified their usurpation of Indian lands. (p. 12)

Colonialists cleared forests at alarming rates, both to provide "farmland and fuel" for themselves, and also to deprive Native Americans of "both forest cover and the sustenance on which they depended for survival" (p. 13). They placed bounties on wild animals and destroyed the ecological habitat, greatly reducing the numbers among animal populations. The sweeping destruction of forestlands and animal life exemplified a mindset of mastery and exploitation of the wilderness toward human ends, an attitude which was apparent in shaping environmental policies and practices for the next two hundred years (Payne, 1996). Two schools of thought commonly understood by 1875 were those of (1) conservation and (2) preservation (Payne, 1996). The former "emphasized the efficient use...and development of material resources" (p. 3) with a human-centered focus toward instrumental ends. The first attitudes toward the environment were "conservation" measures, suggestive of perspectives that viewed the natural world through an anthropocentric lens. Natural resources were valued for their use rather than for their own inherent worth. However, through the nineteenth century increasing regard and appreciation for the importance of preserving the biological world began to mount. The second primary outlook on the environment from the late 1800s, "preservation," assumed a life-centered, or biocentric perspective . . .tak[ing] factors other than human needs and desires into account when making decisions that affect the environment..." (p. 3). Less anthropocentric, preservationists held that nature should be protected for its own intrinsic value and not for instrumental purposes. Preservationists began to argue for human steward-ship of the natural world, not only for the sake of human beings and the generations that would follow, but simply because other forms of life also had rights to live and thrive in their own ways and for their own sake. This perspective came to affect mass opinion and public policy in significant ways. "Nature writers" such as Ralph Waldo Emerson, Henry David Thoreau, John Burroughs, John Muir, and Rachel Carson contributed to a growing awareness of the importance of the natural world. Today, conservation is seen by many as a "politically conservative, anthropocentric approach to environmental protection," (p. 6). The conservationists' focus toward material gains sought through the control and management of environmental resources is gradually diminishing and being replaced by a respect for the intrinsic value of the natural world, resulting in a rise in support for preservationist perspectives among many environmentalists today. In the late twentieth-century, conservationism has given way to "preservationism—with its emphasis on moral, spiritual, aesthetic, and biocentric rationales for environmental protection" (p. 6).

The domination of nature is really about the domination of human beings, says William Leiss (1972). Technological advances made possible by the exploitation of nature, he says, "enhance the power of ruling groups within societies and...as long as there are wide disparities in the distribution of power...technology will function as an instrument of domination" (p. 121). Carolyn Merchant (1996) says that "problems of ecological deterioration, depletion, and pollution...are intimately connected to multinational capitalism," in its wide reaching effects resulting in "the greenhouse effect,...ozone depletion,...deforestation,...pollution,...and species extinctions (p. 187). According to Bowers (1995), "[t]he ecological crisis is, in part, a crisis in values and beliefs" (p. 2). For example, the "myth of social progress...[which] is predicated on an anthropocentric view of the universe" fuels the belief that "our rationally-based technology will always enable us to overcome the breakdowns and shortages" that come with a primary focus on "human interest and technological empowerment" (p. 4). Ecological concern "becomes part of a social movement" (Leiss, 1972, p. 22) because ecological destruction is bound up within cultural beliefs and practices supporting multinational capitalism. Therefore, reversing current trends of environmental degradation will require "challenging the authoritarian decision-making powers vested in corporate and governmental institutions" (p. 22). The domination of nature goes hand-in-hand with science and the advancement of technology. Going back to Bacon, Leiss explains that

> any critical examination of the idea of mastery over nature must confront the thesis that has shaped the common understanding of this notion for several centuries: the conquest of nature by man is achieved by means of science and technology. (p. 101)

As discussed in chapter 1, science turned *scientism* is a process of instrumentalization whereby humans with a consciousness of patriarchal domination exploit nature to their own self-serving ends. Heidegger explains human subjugation to scientific rationalism as

> the will to dominate...at work in any rational discussion or enterprise...a way of questioning things by which they are reduced to enslavement. (as cited in Prigogene & Stengers, 1984, pp. 32–33)

It is in "the logic of capitalist colonialism" that nature should be "appropriated, preserved enslaved, exalted, or otherwise made fixable for disposal by culture" (Haraway, 1991, p. 198). Salleh (1997) questions the U.S. assessment of itself as the "leading nation on earth" (p. x). Failing to put post–World War II "military production to good civilian use," the United States is not a people's republic, she says, "but a welfare system for the brotherhood in suits who direct a complex of tele-pharmo-nuclear corporations" (p. x). She asks the question: "Is there a subject whose labor, and therefore political sensibility, is not implicated in industrialism and its parcels of administered time? Who is equipped to design an ethical constellation that is workable beyond commodity production" (p. x)? She argues that ecofeminism is in such a position. As a border politics, ecofeminism combines "socialism, feminism and ecology" in a dialectical zig-zag approach, recognizing that its triangulation makes it impossible to "go after its political object in a simple, linear way" (p. 108). Specifically,

> ecofeminism moves back and forth...between (1) the liberal-socialist feminist task of arguing its equal right to a political voice; (2) a radical poststructuralist feminist task of deconstructing the masculinist biases of that same political validation; (3) pursuit of its ecological aims by narrating how women have been able to live an alternative relation to nature from men and how men might join them in this way of being. (p. 108)

In looking at "Man's relation to Nature," the ecofeminist project combines those of environmental "ethics and politics" with a simultaneous critique of "Man's relation to Man," drawing on the "socialist project," and also examining "Man's relation to Woman," using critiques from feminist standpoint theory (p. 108). In that women's labor in both First and Third World countries "provide...the largely invisible social infrastructure that mediates between nature and men's economic production" (Merchant, 1996, p. 205), an inquiry into the question of the Human/Nature relation would be remiss without incorporating critiques of androcentrism undergirding human institutions, beliefs, and practices.

There is need to realign our androcentric and anthropocentric ways of knowing within more ecocentered worldviews. A joint statement made to the UN General Assembly in 1993, signed by "1600 scientists from 70 countries, including 100 recipients of the Nobel Prize" (Bowers, 1995, p. 4), gives the following warning:

> A great change in our stewardship of the earth and life on it is required if vast human misery is to be avoided and our global home on this planet is not to be irretrievably mutilated. The Earth is finite. Its ability to absorb wastes and

destructive effluent is finite. Its ability to provide food and energy is finite. And we are fast approaching the Earth's limits. (1993, p. 2; as cited in Bowers, 1995, p. 4)

The scientists at the meeting that produced this document predicted that a mere forty years' time is left for humans to change "destructive cultural values and practices" before we will reach "critical thresholds in the life-sustaining capacities of natural systems" (p. 4). In order to reverse the inevitable momentum that appears to be mounting, it will be necessary for "the North...[to] review its high tech consumption in favor of more species-egalitarian models by which the South provisions itself for the sake of global justice and sustainability" (Salleh, 1997, p. 111).

The reduction of nature to an object for human use is analogous to ways that patriarchy has often objectified women in instrumental terms, reflective of the hierarchical, power-over mindset undergirding mainstream Western culture for most of its history (see chapter 1). Although feminist theory cannot be reduced to a set of common beliefs, most feminists would agree that a patriarchal view has become the preeminent epistemological frame over the long tenure of Western, white-male power-holders. Spretnak (1990, 1993) suggests that the spiritual dimension of ecofeminism provides an alternative to the Western patriarchal worldview of fragmentation, alienation, agonistic dualisms, and exploitive dynamics. Ecofeminist scholars have offered much in the way of rethinking our understanding of language and its effects on human relationships, suggesting, among other things, the limitation of the dualistic models of the patriarchal mindset.

Ecofeminist Riane Eisler (1987) describes a patriarchal view as a "power-over" dominator model because it is one of "ranking" rather than "linking" (p. xvii). In delineating patriarchal from matriarchal views, many ecofeminists make precisely this distinction: a patriarchal view is one which sees the world in terms of higher or lower, whereas, a matriarchal view is one which looks at the world in relational ways. She describes the "dominator model"—whether enacted by males or females—as "the *ranking* of one half of humanity over the other," unlike the principal of linking which "eliminates notions of rank" (p. xvii) and characterizes a *partnership* model. By using these terms, I do not mean to suggest that men, in general, see the world hierarchically and women, in general, see the world in terms of relationship. This exaggeration essentializes men and women as if each gender had a unitary view collapsible into one category or the other. I know lots of men who appear to think in relational ways, whereas, I often encounter women who respond to me competitively or from a vertical position, as if they were above me. Eric Fromm (1986) says that the salience of analogies describing patri- versus matri-focal views is in their metaphorical value, and that such a discussion is less about particularities of women versus men, per se, than it is a naming of *worldviews* as being relationally based or those which are hierarchical (p. 104).

In moving away from the patriarchal *"power-over"* attitude controlling societal relations, feminist author and activist Starhawk suggests a theoretical frame combining *"power-from-within"* and *"power-with"* as the "full sense of genuine linking partnership" associated with matrilineal ways of relating (as cited in Sky, 1993, p. 9). Such a move constitutes a shift from a vertical frame to a more horizontal one by which to view the Other, and is a "re-visioning" which carries with it seeds toward more egalitarian social relations (p. 10). One result could be a "significant lessening of human-caused abuse" (p. 10) often justified as "God's will," or for others' "own good," or that it is "biologically ordained"—which are all examples of the "doctrines of patriarchy" (p. 10). The way we as a species structure our male/female relations, according to Eisler (1990), has major effects on the entire "social system." Perspectives on gender relations connote "individual roles and life choices of both women and men," and also "whether a society will be peaceful or warlike, generally egalitarian or authoritarian, and living in harmony with or bent on the conquest of the environment"(p. 26).

A partnership model is thought by many (Collard, 1989; Eisler, 1990; Goodrich, 1989) to be the prevalent structure within prehistoric matrilineal societies, recognizing human relations as more horizontal than vertical, and one in which a sense of communal sharing was the model for living. Such societies can be traced back to Paleolithic times, the period beginning 25,000 years ago and considered to be the start of Western culture (Eisler, 1990, p. 24). Extensive archaeological exploration has revealed that these non-male-dominated societies lived peacefully—women and men in partnership with each other and with nature—thousands of years before the so-called "cradle of civilization in Sumer" (Mellaart, 1987; as cited in Eisler, 1990, p. 25). Ruether (1996) states that the shift from "egalitarian classless societies" came about with

> a series of invasions by patriarchal pastoralists from the Northern steppes sometime in the sixth through third millennia B.C..E. in the ancient Middle East, reshaping earlier egalitarian societies into those of militarized domination. (p. 4)

Prehistoric clans who worshiped the Goddess are thought to have been matrilineal clans, according to Andree Collard (1989). With invaders came the "social system of war, violence, and male domination," Ruether (1996) says, along with a "concept of God as patriarchal warrior and ruler, outside of and disconnected with nature" (p. 4). The new image of God was very different from one held by the earlier clans, where all on the earth were connected by the life force of a nurturing female spirit. Collard (1989) notes that "when men invented their gods, they projected onto them isolated individualism, hierarchical relationships, and power-based values" (p. 8). Matriarchal societies were based on "kinship, egalitarianism, and nuturance-based values" (p. 8) projected not only toward their Goddess but also toward other species in the environmental surroundings. The perception that women were closer to nature, she says, was an initial link toward the personification of a divine creator as female. Ruether (1996) points out how women often were and

> remain the primary food gatherers, the inventors of agriculture. Their bodies are in mysterious tune with the cycles of the moon and the tides of the sea. It was by experiencing women as the life-givers, the birthers of children, the food-providers, that early humans made the image of the female the first personification of the divine,…source of all life. (p. 4)

In archaeological findings carved symbols of "large-hipped, often pregnant…Venus figurines" (Eisler, 1990, p. 24) have been found in Paleolithic caves. These carved symbols are determined to be the

precursors of the Great Goddess still revered in historic times as Isis in Egypt,
Ishtar in Canaan, Demeter in Greece, and later, as the Magna Mater of Rome
and the Catholic Virgin Mary, the Mother of God. (p. 24)

The relational worldview of Goddess-worshiping cultures is supported
by the notion of "Gaia," Greek for earth, a term applied by biologists Lynn
Margulis and James Lovelock (1989) to their theory that the Earth is a
"living system designed to maintain and to nurture life" (as cited in Eisler,
1990, p. 26). The Gaia hypothesis lends a scientific intersection with the
Goddess-worshiping beliefs of prehistoric societies viewing the world as
a "great Mother...who creates and nurtures all forms of life" (p. 26).
Indeed, tribal lore of current times shows revealing evidence of a respect
for "the unity of all life" and a reverence "for the Earth as our Mother" (p.
26). As humans made a move from being hunter-gatherers to agricultural-
ists, both the increase of "experiences of the animal and plant worlds,"
along with an expanding consciousness, gradually helped to widen their
view of the Goddess who came to be perceived within the surrounding
environment of plants and animals, as well as continuing to dwell in more
ancient forms. Earth-based worship, says Collard (1989) sometimes
referred to as "animism," is a belief that everything that lives is endowed
with Soul/Spirit, recognizing value and offering respect to every form of
life (p. 9). She says that the idea of Goddess worship is important because
women benefit by having a knowledge of their own history as it existed
before patriarchal oppression. She says that women with some "vision of
what we were" will be better able to imagine "what we can be" (p. 8).
Collard discourages women's "incorporation into man's world on an
'equality' basis, meaning that woman absorbs his ideologies, myths,
history, etc. and loses all grounding in her own traditions" (p. 8).

Worship has been defined as the "reveren[ce] accorded a deity...or a
sacred object;...adoration...[as] profound love or regard" (*American
Heritage Dictionary*, 1996, p. 2059). From the root *wer*, meaning worth, it
can be linked to its derivative, "breath," which is a direct etymological
descendant of the term "spirit" (p. 2132). Spirit has been culturally
constructed through language, myths, and metanarratives, named and
worshiped in diverse forms within multitudes of cultural contexts. The

names we give Spirit, in my opinion, are less important than its cultivation as an experiential awareness or sensibility for knowing and being in the world. Whether one inscribes it as a Goddess or a God, a savior or a saint, that animating "breath" of life, that *elan vital*, is one of the offerings of the natural sensate world of human experience. As a human, I am as able to draw upon its sustenance as I am able to draw air into my lungs. Some of the ways in which that sustenance comes to me is in the form of wonder, awe, joy, gratefulness, so that the experience of its magnitude imbues my life with a color and richness that is difficult to articulate.

Moon Magic

The moon outside was clear and full tonight, lighting up the few small clouds hanging low over the woods. It shimmered down onto a stretch of grass behind the cabin—clinging particles of dew caught its light, the prisms of a thousand fireflies. I awoke from dreams—an awakening into magic—I could feel it, so remarkable. I moved into the darkened kitchen and sat upon the floor before the window, the illumined, spherical object/not-object high above me. I could feel its intense radiance penetrate the windowpane. And there in the embrace of its light, I was *with* it and it *with* me; its beauty and its power, so full, and I mirroring that radiance back, a joyous welling up, my gratitude.

Perhaps that is what worship's all about and why it's so desirable to God. The mirroring back of grace, of love, of joy—of God's sheer magnitude—is how God knows God, experiences her/himself. The moon is Other and not-Other: its light is beauty in my gaze. And my gaze, not of disconnection, separation, the one-way transmission of a spectator who sees but is not seen, takes but gives nothing in return, never in participation.

What is this then, this glowing, this emanating-back, if it is *not* worship? And what of the space between? The moon's effects on me would be little without the radiance of its light traveling across the distance that "separates" us. And across to me I sense some response, some awareness of the mirror I'm providing. In the *between* we come together, and there within a nexus of relations something altogether new emerges: creation.

Never-Final Words on Difference

Ecofeminism, according to Merchant (1996), does not presume nature to be "necessarily a sphere of harmony and peace" where women won't

> be in conflict or manipulate to their own advantage. Nor does it raise feminism or woman-centered culture to the forefront as a way of moving, beyond dualism. Rather, it redefines reproduction as involving powerful forms of creativity and knowledge that are positioned in alliance with nature rather than against it. (p. 204)

In recognizing there will always be decisions of responsibility based on difference, Lyotard (1979) speaks of rival factions within different "language games." In cases where there is no criterion by which to adjudicate because procedural rules of each are incommensurable, he calls them differends (pp. 65–67). He gives the example of two people, one who claims to own an apple tree, the other having harvested the fruit and cared for the tree, but who does not believe that apple trees *can* be owned. In this case, deciding in favor of either will "wrong" the other because the criterion by which to reach a rational decision is different for each. Their situations are incommensurable (p. 66). Within these games of language, he says that differends will "inevitably arise" because understanding across phrase regimens is incommensurable. In adjudicating this conflict between "differends," he says, the most we can hope is to do the *least* amount of harm, knowing that in legislating between incommensurable vocabularies, choosing one will always "wrong" the other. Still we must act, and so we try and do the least amount of harm and name and account for the incommensurabilities instead of veiling our inadequacies in illusions that we may always attain "the good" or reach the horizon of consensus (Lyotard, 1979). Sometimes the best that we can do is to agree to disagree.

A view on difference and relationality, however, can be seen as a continuum or a confluence of forces rather than a binary, according to Val Plumwood (1996). She says that the

> view of self-in-relationship…avoids atomism but . . .enables a recognition of interdependence and relationship without falling into problems of indistinguishability, that acknowledges both continuity and difference . . .it

bypasses both masculine "separation" and traditional-feminine "merger" accounts of the self...provid[ing] an appropriate foundation for an ethic of connectedness and caring for others. (p. 172)

Such a re-vision of difference requires a shift from an ego-centered view to one which is eco-centered, and is necessary for an ecospiritual ethic to which we now turn.

Notes

---

1. Defined by Karen Warren (1993) as "the unjustified domination, exploitation or destruction of nonhuman nature" (p. 122).

2. Discussed below.

Chapter Four

> Our love and attention and devotion to the earth is always worked out in small and meticulous ways...ecological and pedagogical responsibility comes to bear here and here and here, in this next gesture...It is borne out in how we live our lives on this precious earth.
>
> —David Jardine, 2000

An Ecospiritual Ethic and Decisions of Responsibility

I awaken to the sound of a call—"bob white"— just outside the window where I sleep. My eyes open with its solitary distinction, at once it is both low and shrill against the morning's quiet. My awareness sharpens. Further down the bayou, a screech owl moans its high cat-like "awl." I distinguish the sound. One at a time there begin to come others—the soft trill of a morning dove, the call of a cardinal and then a thrush, the voices begin to rise and build into a multiplicity of song. How have I missed this before? Was it my post-dream consciousness that brought the world into such sharpness? The images come on clear and startling, immediate. It is as if there is some register within myself that pulses with each tone, which opens and connects as an affinity between and across forms, a movement bridging a wide span of differences. It is an opening door between myself and "the Other," a translation across differing vocabularies into some instant of

> 'becoming' which is not common to the two, since they have nothing to do with one another, but which is between the two, which has its own direction, a single bloc of becoming, an a-parallel evolution. (Deleuze, 1977/1987, p. 7)

An "a-parallel evolution" is not an instance of 'one-becoming-the-other' or even 'becoming like the other,' not reduction nor imitation, for the experience is unique to each entity involved. Instead, it is "something which is between the two, outside the two, and which flows in another direction" (p. 7), something which arises out of difference, fresh with the generativity of new creation.

The Buddha taught that there is no "*a priori* reasoning, no realm of pure logic aloof from or unconditioned by the sensory world" (Macy, 1991a, p. 67). He said that perception is a product of three factors,

> a sense organ,…a sense object coming within its range, and…contact between the two. These conditions…constitute the gateway through which perception occurs…subject and object are interdependent. (pp. 67–68)

It is the *interdependence* of one and the "Other"—or many others—that I will address in this chapter, holistic and ecologically informed relational ways of knowing found within theories of curriculum, ecofeminism, and deep ecology, leading to an ecospiritual ethic. Geologian Thomas Berry (1988) has said that "one of the historical roles now being assigned to our generation is the role of creating…the spiritual context of the ecological age," which he says at this point in history is "presently taking on its effective form" (p. 119).

I am interested in how spirituality may contribute to a re-vitalization of the human relationship within the ecological world, a relation enhanced by a sense of the *sacred*. An ecological ethic is an eco-centric approach to living that "places humans within, rather than above or outside of nature…and places value on all entites that are self-renewing,…from individual organisms to the ecosphere itself" (Merchant,1996, p. 205). Robyn Eckersley (1992) says that ecofeminism is one of "several varieties of ecocentrism," which values "not just individual living organisms, but also ecological entities at different levels of aggregation, such as populations, species, ecosystems, and the ecosphere (or Gaia)" (p. 47; as cited in Merchant, 1996, p. 204). An "ethic of earthcare" is Patsy Hallen's (1991) "comprehensive vision for a new worldview with ecofeminism, feminist science, and process philosophy as its core components" (Merchant, 1996, p. 206). Hallen brings together women exemplifying "alternative scientific and philosophical traditions," synthesizing their approaches and applying them "to concrete situations, such as saving Australia's ancient forests"

(Merchant, 1996, p. 206). An analysis of gender and other forms of oppression will assist in shifting our ways of knowing from anthropocentric to eco-centric ones because a patriarchal basis in separation underlies racism, classism, and naturism. Therefore, in addition to a re-vitalization of the concept of relational thinking, there will be an accompanying emphasis on the importance of difference.

An ecospiritual ethic includes alternative ways of viewing the world as heuristics toward a re-conceptualization of epistemological practices—for people both inside and outside of education—and a recognition of the deep and long-term connection humans violate by destroying the earth. It seeks to initiate a consciousness-raising among people so to foster an emerging ecological intelligence as part of humans' evolutionary process. This eco-intelligence includes: a "rethinking of the nature of intelligence" (Spretnak, 1997, p. 122) away from the "model of the detached spectator" (p. 122), a re-evaluation of the ideology of individualism underlying anthropocentric worldviews (Bowers, 1995), and a recognition of the value of "place and responsibility" as part of an interdependent web of life-process. Belenky et al. (1996) remind us of the beauty of the metaphor of a "web" to describe the life-world: "In the complexity of a web, no one position dominates over the rest. Each person—no matter how small—has some potential for power...[and] each is always subject to the actions of others" (p. 178). An ecospiritual ethic will require ontological work toward a way of living that is viewed as a constant state of "becoming," a procession of negotiation and change shaped by, and also shaping, our constantly changing ways of knowing and of being in the world. Thus, an ecospiritual ethic informs and is informed by an ecospiritual *praxis* entailing a continuous reflection on who we are, and on who we wish to become.

Necessary to this perspective is a willingness to critically reflect—both personally and culturally—on where we've been, in relation to where we are and to where we want to go. With reflection must come a willingness to deconstruct the prevailing beliefs and practices (Salleh, 1997) that alienate human beings from each other and from the environment with little regard for the relationships holding us in an embrace of symbiotic balance. "Changing the foundations of taken-for-granted beliefs," Bowers (1995) says, will require, not only a "focus of attention," (p. 6), but decisive efforts to see that reforms are carried out. Change will demand a conscious effort of humans to shift their "natural attitude" away from a "con-sumer/technologically driven life style" and toward creating a natural sensibility that supports "ecologically sustainable cultural practices" (p. 6). Pressing areas for change include: "Reestablishing climate stability, protecting the...ozone layer, restoring the earth's tree cover, stabilizing soils, safeguarding the Earth's biological diversity, and restoring the traditional balance between births and deaths" (Worldwatch Institute, 1993, p. 17; as cited in Bowers, 1995, p. 6). According to Berry (1988) "what is needed...is the deeper meaning of the relationship between the human community and the earth process" (p. 10) so that we may cultivate "our sense of gratitude, our willingness to recognize the sacred character of habitat, our capacity for the awesome, for the numinous quality of every earthly reality" (p. 2).

Matthew Fox (1995) has spoken of an inherently nondualistic "postmodern spirituality" that combines praxis with reflection on that practice" and "develop[s]...the powers of creativity, justice and compas-sion in all persons...as the basis for a community's rebirth" (pp. 5–7). Such a rebirth is part of the cosmological process that undergirds life at every level and is reflected in the surrounding environment. A cosmological view does not privilege one person over another according to gender, race, or class. It recognizes the dynamic interplay between all forms, human and nonhuman, as contributing to a process which is fundamentally based on diversity. Fox (1983) has described the cosmological view as evoking "a sense of balance, of harmony, and therefore of justice" (p. 70). He explains that "the word 'cosmos' is in fact the Greek word for 'order.' A cosmic spirituality is a justice spirituality, for it cares with a heartfelt caring for harmony, balance and justice" (p. 70). An awareness of the human's place within a larger cosmological order can be cultivated, in part, through attention to increasing the individual's powers of personal reflection and action toward change, a personal praxis that attends to the living of life. This actively engaged ecospiritual praxis places the individual as a respon-sible participant within the experience of his/her own "becoming," as a never-ending process of negotiation within interdependent nests of relations (Jardine, 2000).

The idea of interrelationship is important to educational theory. The word "whole" designates "a full amount...not divided or disjoined...an entity or system made up of interrelated parts" (*American Heritage Dictionary*, 1996, p. 2038). David Bohm (1985) notes that parts are not the same as fragments. If I were to smash a watch on the table with a hammer, I would produce fragments, not parts, because "they would no longer be significantly related to the structure of the watch" (p. 23). Furthermore, in constituting a whole as I define it, parts are generative in their synthesis, producing more than the sum of which they're comprised. The term "whole" comes from the Old English *hal,* some of its derivatives being, "health, heal, holy," i.e., "sacred," and "hallow," which means to "bless" or to "consecrate" (*American Heritage Dictionary*, 1996, p. 2038).

Curriculum scholars historically have encouraged the incorporation of ideas considered characteristic of more holistic approaches to curriculum and schooling (for example, Dewey, 1956; Greene, 1995; Huebner, 1999; Macdonald, 1964/1995; Zirbes, 1934), such as the need to integrate content areas across the curriculum and to emphasize a sense of community among the people involved in the processes of education. Many have generated discussions about how to widen and deepen our ways of educating children more holistically—as physical, emotional, mental, and spiritual beings (see Pinar et al., 1995). Unfortunately, the fruitfulness of these discussions has been limited, in part, by the imposition of Western conceptions of relationality that have been hierarchically ordered according to patriarchal worldviews framed in dualistic thinking.

A hundred years ago, John Dewey (1900,1902) offered the world a vision of education which emphasized personal fulfillment and social well-being within an actively experiential context that "was to include the aesthetic, the contemplative, and what some would call the spiritual aspects of human experience" (Jackson, 1956; as cited in Dewey, 1956, p. xxxvi). Yet one major stumbling block, which Dewey continually sought to disrupt, was the dualistic thinking of a culture that continued to pose school against society, or child against curriculum (1902). This either/or attitude, so prevalent in Western culture, separates the dimensions of human potential into categories and linguistically "codes" them as *either* important *or* unimportant in educational terms. Typically, what is valued are those areas assigned to the cognitive domain rather than those relating to the physical or affective.

Binary thinking is an important analytical tool that has given humans scientific and technological abilities that continue to improve the world in countless ways. Yet, Parker Palmer (1998) suggests that our "Western commitment to thinking in polarities…fragment[ing] reality into an endless series of either-ors" has become so pervasive that we "think the world apart" to the extent that we "destroy the wholeness and wonder of life" (p. 62). He says that seeking "truth" only through the distancing and reductionist methods of binary logic is inappropriate in areas where both/and thinking would better serve understanding. This way of knowing and experiencing the world would assist in helping people "think the world together" (p. 62) toward "develop[ing] a more capacious habit of mind that supports the capacity for connectedness" (p. 62) fundamental for engendering more holistically based cultural practices.

Belenky et al. (1986/1997) support the *integration* of dichotomous binaries (both/and thinking) in their studies delineating categories of knowing found in populations of women of different age, class, and ethnic backgrounds. They describe a type of "reasoned" knowing they term "procedural" because it follows methodological procedures for reaching conclusions based on evidence. Within this category of "rational" thought, they include not only the objective, distancing posture considered valid in scientific research ("separate knowing") but also a relational method in which the knower attempts to understand the object from an empathetic or a "thinking-with" perspective ("connected knowing"). Both separate knowing and connected knowing are procedural ways of distancing the knower in order to better understand a particular case under study. Each follows a method based on reasons valid to the knower. In the former the knower stands back to observe analytically, while in the latter the knower separates from her/himself in order to connect with, and observe from, the perspective of the "Other."

The authors noticed that when women who were "procedural knowers" could integrate both separate knowing and connected knowing, they were able to complexify the relationship of knower to known as a foundation for an awareness that *"all knowledge is constructed,* and *the knower is an intimate part of the known"* (p. 137; emphasis theirs). Many who demonstrated this sort of integrative ability to *both* "think the world apart" *and* "to think the world together" (Palmer, 1998, p. 62) were learning a means for "weaving their passions and intellectual life into some recognizable whole" (Belenky et al., 1986, p. 141). As these "constructivist" women developed a more integrative way of knowing, they experienced new ways of interfacing with their worlds. For example, they made a distinction between what they referred to as "really talking" as compared with a more "didactic" means of communication. The latter was more a type of "hold[ing] forth" than was experienced in the former that they described as a sharing of ideas (p.144). "Real talk," similar to what Kohli (1984a) calls a *"genuine exchange"* (p. 36), is a possibility sometimes reached through dialogue that is initiated from a position of "equal footing, equal grounding for conversation and for understanding" (p. 36).

This type of dialogue is not always possible, or achievable, but I would suggest that the dialogical reciprocity Kohli describes is supported by a way of knowing that is open and willing to recognize, indeed listen to, the "Other." Palmer (1983) asserts that "the crucial difference between observing and relating is that a relationship is always two-way" (p. 54). To operate only from the standpoint of "objectivity," he suggests, is a "limited mode of knowing" (p. 54) because it denies the voice of the "Other." Palmer says that through one-way observation, "not only does the nonhuman world remain inarticulate, but the human world is deprived of its voice as we transform people into objects, things" (p. 54). Using a variety of our capacities for knowing, such as "empathy, intuition, compassion, faith," he says, allows us to

> pick up the world's subtle signals, its subvocal speech, its messages to us about our limitations and responsibilities and potentials. When we allow the whole self to know in relationship, we come into a community of mutual knowing in which we will be transformed even as we transform. (p. 54)

Narrowing our view of the world through a reductionist lens of either/or thinking has its place in our knowing, yet to stop there imposes limitations on potentials for human thinking and being in the world. It should be noted that the models Belenky et al. delineate are not present in a pure form within individuals, rather, there is overlap and combination across categories of knowing within particular people. A participant in their study who was a "senior honors student in science" clearly stated both the benefits and the limitations of this sort of study by saying that

> in science you don't really want to say that something's true. You realize that you're dealing with a model…[which is] always simpler than the real world. The real world is more complex than anything we can create. We're simplifying everything so that we can work with it, but the thing is really more complex. When you try to describe things, you're leaving the truth because you're oversimplifying. (p. 138)

In that these studies were done on populations of women, their relevance to men's thinking can only be speculated. Yet, Belenky et al.'s models help complexify our understanding of ways that *how we know* shapes how we construct the world around us, even as the world in which we live is simultaneously constructing us. The task of developing the rational mind has overshadowed work on other equally important areas of human potential. Susie Gablik (1991) points out that an

> insistence on the relational nature of reality is precisely what is missing in the Cartesian paradigm and it would seem that what we are beginning to experience, at the leading edges of our culture, is the dismantling of Cartesianism—the paradigm of the bipolar subject and object. (p. 164)

To know is always reciprocal—knowing is a two-way relation. That understanding is reinforced by areas of spiritual education that have begun to blur the boundaries long dividing human understanding into categories of absolute terms (Smith, 1996). And with that understanding, we now move to a discussion of relational ways of knowing and of being in the world.

Increasingly, Buddhism is proving useful for scholars interested in making a shift "toward a dynamic, systemic, process view of reality" (p. xi). David G. Smith (1996) draws on the notion of Buddha-dharma, which in the original Sanskrit means "one who is awake" and also "carrying" or "holding," respectively. He says that

> studying the Buddha-dharma, then, refers to the action of being awake to, or attending to what carries, upholds or sustains us as human beings. (p. 8)

Dharma-consciousness is a recognition of the deep sustenance that comes from an awareness of the human connection within a larger web of relationships. The network of relations extends from within the self (the personal) to "the other" (the community) to the wider planet and its universal framework of connection. Implicit within the idea of Buddha-dharma is the systemic nature of life, a cosmological view that all reality is "dependently co-arising" within a web of relationships which are multidimensional" (Macy, 1991b, p. xv). This idea is exemplified by the story of Indra's jeweled skirt, a

> multidimensional net [that] stretches through all space and time connecting an infinite number of jewels in the universe. Each jewel is infinitely multifaceted and reflects every other jewel in the net. There is nothing outside the Net and nothing which does not reverberate its presence throughout the web of relationships. (Kaza, 1993, p. 57)

Joanna Macy (1991b) is an ecofeminist who describes a systemic framework for living wherein "each and every act is understood to have an effect on the larger web of life, and the process of development [is]...perceived as multidimensional" (p. xv). Interdependence, in Macy's estimation, suggests "these developments" are not sequential in a linear way, but each occur "synchronously...reinforcing the other through multiplicities of context in which other events occur" (p. xv). She traces some intersections between fundamentals of Buddha-dharma consciousness and postmodern process theory. The latter draws on biological systems theory (Doll, 1993; Varela, Thompson, & Rosch, 1991) as an analytic for understanding all forms—human and non-human—as varied levels of interlocking organizations existing at different levels of aggregation. According to Doll (1993), "an open, biologically oriented" systemic model stresses "interaction." He says that

> in a living system, parts are defined not in isolation from one another but in terms of their relations with each other and with the system as a whole... mak[ing] it...more appropriate as a model for human development and categorically different from Newtonian physics. (p. 64)

In looking at a community as a living system, for example, one can see that a community is not a bounded, self-contained unit, suggested by the symbol of a circle. Rather, it has certain constraining parameters that are also balanced by openness and interchange. Boundaries are not fixed or static, so that a shift in perspective across boundaries reveals wider dimensions of combination or intersection with other systems (Laszlo, 1972) in broader and more all-inclusive ways. An image of intersecting circles as in a Venn diagram might symbolize more appropriately the ways that a community is in reality a multiple construct, overlapping levels-within-levels, cooperating, contradicting, coinciding, misaligning, so that "community" is a complex and contradictory form, as are living systems.

Complexity is one of the characteristics of a biological open-systems view. Doll (1993) provides an example of this part-to-whole relation by describing the "relationship between cells that constitute the brain and the brain as a functioning whole system" (p. 65). At the more basic level of aggregation, the brain cells are "a ceaseless change of detail," while overall, "our behaviors, our memories, our sense of integral existence as individuals retain" a unified pattern of organization (p. 66). Thus, the brain is "chaotic" at one level, while "at another it is complexly patterned" (p. 66). And in the final analysis, he says,

> [t]hese two perspectives cannot be substituted for one another, nor reduced to one another; instead, they are complementary, indeed integrated. (p. 66)

This view of an "interconnected cosmic web," according to physicist Fritjof Capra (1975/1991), has been used in Eastern philosophy and religions to "convey the mystical experience of nature" (p. 139). Within interlocking systems, "self, society, and world are reciprocally modified by their interaction, as they form relationships and are in turn conditioned by them" (Macy, 1991a, p. 99) a fundamental principle within the Buddhist idea of dependent co-arising.

As well as systems theory, a process view of reality that is consonant with Buddhist philosophy includes the idea of an inherent order exemplified by patternings which can be seen across different levels of aggregation. Gregory Bateson (1979) draws on the notion of multidimensionality as a useful heuristic to view all phenomena through a relational frame of connecting patterns (see also Capra, 1975/1991). For example, planetary bodies being held in orbit by the constraining force of gravity exemplify such a pattern: the pattern created by planets in orbit is also replicated in a very similar form, but at quite a different scale, or level of aggregation (Hallen, 1991) through the lens of a microscopic camera inside the human cell. When one begins to look for the patterns that connect, one will find no thing in isolation. The idea of an underlying order to all matter is one supported by physicist David Bohm's (1985) discussion of an "implicate order" underlying all phenomena (p. 18). According to Capra (1975/1991), the metaphor Bohm uses to exemplify "this implicate order" is that of a "hologram," because of its property that each of its parts, in some sense, contains the whole" (p. 320). Bohm (1985) explains that similar to a holograph,

> the whole universe is in principle enfolded into each part...[so that] each part is in a fundamental sense internally related in its basic activities to the whole and to all the other parts. The mechanistic idea of external relation as fundamental is therefore denied. (p. 13)

The idea of wholeness[1] doesn't suggest that humans will come to develop an ultimate understanding of all things, but, rather, that reality can be understood "as an unbroken and seamless whole in which relatively autonomous objects and forms emerge" (p. 21). Such a view is suggestive of a part-to-whole relation (Capra, 1975/1991), having corollaries within quantum physics and depicting "the universal not as a collection of physical objects, but rather as a complicated web of relations between the various parts of a unified whole" (p. 138). In human terms, this cosmological view recognizes that we are dependent on more than ourselves. Indeed, that we are

> conditioned by and coexist…in dynamic interdependence with all things. Such a cosmology…would reinvigorate the human in an ethic of reflection upon and care for life in its entirety, as the species which can identify the integrity of the whole in the richness of its diverse particularities. (Brown, 1993, p. 136)

The idea of co-emergence (Macy, 1991b; Bateson, 1979) is useful toward a reconception of the linear mindset of a Newtonian worldview. Darwin's theory of evolution has long been viewed as a linear progression of the straight-ahead, ordered, advancement of biological change. For example, as the strong, the smart, the agile are selected, the species differentiate and improve. But more and more is being said of the other side of the coin, i.e., the *context* in which the mutations occur. The receiver of the action/change is as vital to this process as that which does the action/changing. Mutations never occur in a vacuum but as an action or result arising out of some interchange, some negotiation of difference. And mutation occurs because there is something in its environment which is ready for it, which receives it, and is vital to what it becomes. Without the receiver, it may have become something altogether different. As Bateson puts it, the

> messages cease being messages when nobody can read them. The power to create context is the recipient's skill…[this] genesis of the skill to respond to the message is the obverse, the other side of the process of evolution. It is coevolution…[because] it is the recipient of the message who creates the context (p. 48)

In contrast, the one-way orderly progression forward suggests a patriarchal perspective, a power-over frame that negates the equal importance of the receiver, the context, a mutable form. Evolutionary change is not a one-way relation of cause, then effect. Evolutionary change is equally influenced by that which is ready to receive and adapt to the initiating momentum of a said cause. The significance of being on the receiving end of this momentum is mutability itself. Underlying the evolutionary pattern of survival of the fittest is this mutability. Rather than strength or power to override, it is adaptability that lends longevity to lifespans. The ants and the cockroaches have far exceeded the dinosaurs in evolutionary durability (Bateson, 1979). Whether it is the life span of a Japanese beetle, a planet in the solar system, or the quality of a relationship between a man and a woman, survival of the fittest can translate into adaptability. Rather than a uni-directional linearity in which A solely affects B, a cosmological view of relations recognizes B as co-determining A's action, i.e., idea and context are mutually emergent. And a hierarchical worldview that perceives the active force as determining its effect is softening into one that recognizes the infinite possibilities of relations and their capacity for combination. As well, it is one where the recipient of the action, the "B" as it were, is as significant in determining an outcome as is its cause, "A." In this regard, it could be said that the receiver is as important as the sender, thereby countering hierarchical positions exemplary of Newton's cause-effect determinism, as well as, certain "power-over" positions of patriarchy implicit in a self/object, male/female, culture/nature relations.

I return to the notion of spirituality as a practice of "mindfulness," relevant to this discussion as another way of knowing that draws on and emphasizes our connection within a larger matrix of planetary life and seeks to foster a sense of human "engagement" within that living process (Hanh, 1992). The Buddha was asked a question: "Sir, what do you and your monks practice?" He replied, "we sit, we walk, and we eat." The questioner continued, "But sir, everyone sits, walks, and eats," the Buddha told him, "When we sit, we *know* we are sitting. When we walk, we *know* we are walking. When we eat, we *know* we are eating. This is the difference" (Hanh, 1996, p. 19). And indeed, it is.

Mindfulness is an attempt to live in a state of "full appreciation of each moment." This way of "intentional living" is reminiscent of what Smith (1996) describes as "living awake to the way that sustains us" (p. 9), and emphasizes an intentionality toward caring for myriad forms of "others" cohabiting the planet. Mindfulness suggests a way to enter deeply into the present moment and recognize its significance. As Hanh says, "the miracle is not to walk on water…the miracle is to walk on the green earth, dwelling deeply in the present moment and feeling truly alive" (p. 20). I am reminded of Maxine Greene's (1995) existentialist notion of "wide-awakeness"—an awareness of what it is to be in the world…[and the] longing to overcome somnolence and apathy in order to choose to reach beyond" (p. 34). For myself, implicit in this "reaching beyond" is a recognition of the incredible generativity and also the mystery of life processes (Doll, 2002) within which we are a part, as the human species. It implies a fluidity of perspective in order to "shift our gaze from the particular to the interconnected" (p. 14),

> to awaken a way of seeing, a way of living and of consciousness that in fact every human person is capable of…the micro/macrocosmic vision of the world. (Fox, 1983, p. 71)

This way of viewing the world draws on recent developments in science that are allowing us to increase our understanding of the underlying processes that are creatively generating existence. Charlene Spretnak (1997) points out that the new science of complexity has shown that

> properties *emerge* creatively within systems, while chaos theory has shown that nature moves in and out of patterns of self-organization. Nature at large…is now understood to function more like a creative unfolding than a mechanistic play of stimulus and response…contribut[ing] to a new understanding of our context as a dynamic community. (p. 3)

Unlike this process-oriented view, the "Is" of Western tradition (Deleuze 1977/1987) is an *a priori*, static form as final and complete. Such a view is absolute with no open ground for generative birthings, propagations, subsequent arisings of a-parallel formings; nor is it dynamic or open to as-yet unmanifest possibilities that exist always at the edge of our knowing. The Whole of which I speak is unlimited, unfinished, incomplete, in-process. It is always a part of some wider network of relations and as unfinished as the cycles of life which, even in their withering death, disseminate seeds and seedlings in an ever-renewing procession of generative creating.

A large fig tree spreads across the back of the cabin just outside the window where I write. I watch the birds beyond the glass as they move through the broad velvety leaves. They come in increasing numbers, actively hopping about, fluffing and preening, and leading me to speculate that they're waiting for the figs to ripen. They move along the twisted branches, inspecting the ripening fruit as it swells day by day and seem to look for the reddish hues to tell them that the figs are ready to eat. Soon, they'll be pecking into the sweet flesh, a bird-and-fig-becoming, an assemblage (Deleuze, 1977/1987) constituted by a particular relation or set of relations. The birds will eat, go on their way, and as seeds are cast, the effects of those relations will continue. From a single seed, tiny root hairs will push forth into the earth seeking sustenance, continuing the momentum. Life happens at the nexus of relations, the nodal points that connect the betweens. In the larger scheme of things, there is importance both in an individual entity or single event, and also in the interface *between* forms, the interrelationality that contributes to make up the whole. The whole would not be what it is without all of the singularities, the particular sites it embodies—the bird-and-fig-becomings—as nexus of relations of which the whole is composed. And it is the cohesion of the 'many' making up the 'one' that constitutes every moment of life as uniquely what it is—multitudes of singularities within an ever-changing totality of relations.

I do not propose the totalizing view reified as the "Is" of Western philosophy. What makes my ecospiritual view of holism different is the idea of openness for *movement* within the ever-dynamic flux of life that keeps things generative, that freshens and renews. When we look at the natural world, dynamic movement is everywhere apparent from the molecules within a cell to the rotation of planetary bodies. The constant negotiation between one form and an "Other" could be described as a continual interplay of dialectical tension which is transactional. I use the term "dialectic" with reservation, fearful of Hegel's final consummation in which either thesis or antithesis sublates the other in a power-over, hierarchical maneuver. This final synthesis, rather than denoting movement, suggests the frozen ultimacy of another stagnant form, a death knell for any continuation of momentum. I don't believe "stagnation" is a place we want to go. In re-framing the notion of "dialectic," I like Salleh's (1997) re-casting of the concept as a zig-zag procession—one position playing off of the next in a continuing motion of negotiation and change toward particular directions. In this way "the recursive moment is never complete, because as we move with it, new historical forces come into play" (p. 38) providing a way to re-think life experience as being in continual *process*.

Abram (1996) points out that "every phenomenon…is potentially expressive…Thus at the most primordial level of sensuous bodily experience, we find ourselves in an expressive, gesturing landscape, in a world that speaks" (p. 81). Foregrounding the processional movement in life is one way to visualize how life experience is given meaning, indeed signified, at the nexus of relations, and in that nodal point is the significance that each part brings to the encounter. A nexus is where entities, ideas, perceptions, *existents* converge that give flight to an intersubjective exchange that constitutes life: A young mama cardinal is building her nest in a cluster of fig leaves in the fork of a limb. Soon she'll have young ones and bring insects and berries for them to eat. She'll make the nest more comfortable by tucking in strands of moss or bits of yarn. She'll protect her young when the cat is about, or spread herself over them when the air is damp or cold. She attempts to accommodate each situation that presents itself, particularly as it manifests in the form of "difference." As we understand it, no language is exchanged, yet there is meaning made at the nexus of relations, at the interface between one and the "Other." There is comfort—then cold—she covers them. There is safety—then danger—she protects them. A process view is a relational way of knowing that recognizes the continual "play" between sameness and difference as the constant movement of adjustment within the surrounding network of relations. It is a pattern reoccurring at many levels. In terms of human communication, the dialectical becomes dialogical, providing—again—for *movement* as negotiation out of stasis. It can manifest as attempts to equilibrate understanding between differing perspectives through negotiation, listening, disagreeing, accommodating, arguing, adjusting. When the willingness for open dialogue with the "Other" stops—freezing the momentum—communication ends and relations break down. When either party is no longer willing to be open to hearing the voice of the "Other" or to taking that voice into account as part of a creative synthesis of views, potential "new worlds" are aborted.

In this representation openness for movement helps prevent the stagnation of any one form, for one of the problems of totalities is the problem of closure leading to a tyranny of the whole (Bernstein, 1991). Openness and movement is not all that's needed to prevent totalization, however, not just *any* movement, willy-nilly. History has taught us to ask ethical questions: Where to move? How to move? Who decides? What *is* "reasonable?" *Whose* knowledge? Open systems don't deny agency, in fact, the movement toward *balancing* the "play" between sameness and difference, toward justice-making, toward equilibration is characteristic of the creative intelligence that continually vitalizes and renews healthy biological systems. To maintain fairness, justice, and minimize harm (Lyotard, 1979), humans use judgment to exercise "control" in various ways—meanwhile risking danger in the totalizing effects of knowledge, itself (Foucault, 1979). As Huebner (1999) has said, knowledge can enrich human life, making

> transparent that which seems opaque...relat[ing] events and phenomena which seem unrelated to the unknowledgeable eye...but knowledge...also corrupts. Knowledge has within the power to enslave, to make one less free, rather than freer, unless the user is fully aware of the disadvantages. (p. 37)

One factor to aid in resisting the enslaving tendencies of knowledge, or anything else, is to remain open to the dynamic movement that a process approach to knowing can afford. With that, we must move with eyes open in order to avoid pathways we have followed in the past due to "visions of the whole." Lyotard (1979) warns that there will not come at last some "reconciliation between language games, and that only the transcendental illusion (that of Hegel) can hope to totalize them into a real unity" (p. 81). He says that the "illusion" of totality has brought a great price in the nineteenth and twentieth centuries, a price of terror paid for the nostalgia of the whole and the one. "Attachment" to the one or the many carries danger that one or the other view will become totalized. So it is with great care that we move toward bridging differences. A wary commitment not to privilege either the "one" or the "many" needs be a vital component of an ecospiritual praxis.

Zen Buddhism offers a metaphor, the "middle way" (Watts, 1957/1989), which suggests that it is in the *between* position wherein identification with one or the other referent (attachment to A or B) can be avoided. The middle path moves between the reference points—those of reason/emotion, self/other, one/many—to draw on more "integrative" ways of knowing that I equate with a cultivation of mindful awareness as a basic practice of living. I contend that such a practice is a *choice* to walk a path with conscious *intention*. As well as personal integration, that choice constitutes becoming part of an interrelational community and suggests the bearing that participants within communities may have: to choose to act "mindfully," rather than living life on a kind of automatic pilot. Being pushed by life in all directions with little thought of the part we play within a larger relationship is also a choice, a choice *not* to choose. Living out of a conscious *intention* suggests that each move in life, each choice, is important; that every moment-by-moment decision matters, because it suggests who we are by means of our own construction. And I would suggest that bringing "intentionality" into day-to-day life as part of a personal praxis contributes to the cultivation of an integrative awareness, drawing on heart, spirit, *and* reason, our capacities for action based on informed judgement.

Yet, there exists a dilemma internal to the idea of intentional living: There are those who have no knowledge that making choices *is an available choice*. Bowers (1995) points out how "individual practices are simultaneously expressions of culture" (p.191) coded within the language that we speak. Language, in turn, reflects epistemological structures of embedded beliefs which shape who individuals come-to-be. People who have been marginalized or who begin with an uneven playing field, may have been "defined out" (Kohli, 1995, p. 108) of discourses that empower personal choice as being one of their options. Discourse embodies the cultural codes in which communities are embedded, and these boundaries inscribed by language narrow the ability to express ideas within predefined parameters. My episteme *thinks me* (Bowers, 1995). Bowers (1993) says that instead of assuming that "atomistic individuals think, communicate, and transform the world in terms of their subjective intentionality, we need to recognize the individual as a social-cultural being" (p. 60) which has been formed by social and linguistic characteristics of the culture. In order to educate in ways that empower students, he says, educators must ensure they recognize the

> powerful role that language plays in influencing thought: its metaphorical nature, the influence of its deep epistemological structures on the pattern of thought, and the political nature of language (i.e., the connection of language and power). (p. 62)

Epistemic structures that shape language make it necessary to unpack ways that *how we know* may designate *who we are* as culturally and socially constructed beings and affect whether choices such as "intentional living" will exist within a person's particular realm of experience.

How might rationality as the discourse of privilege and exclusion be reconceived? Critical theorists have proposed a number of perspectives relying on dialogue and an acknowledgment of differences within a community of inquirers (Bernstein ,1991; Habermas, 1976/1979; Kohli, 1995). Through the vehicle of *communicative* reason, for example, Habermas would "challenge the irrationality of the current society" with the "redemptive" possibilities of a truly rational alternative furthering Weber's notion of "rationalization processes" (as cited in Jay, 1973, p. 61). In doing so, he makes a distinction between two types of "rationalization." The first is the aforementioned "instrumental" form described as "growth of the productive forces and extension of the power of technical control...[wherein], reality is objectified according to general laws" (Wellmer, pp. 246–248; as cited in O'Neill, 1976). "Communicative rationality," moves instead toward a practical consensus and "mutual understanding" which "would signify processes of emancipation and individuation as well as the extension of communication free of domination" (p. 246).

Kohli (1995) further complexifies the notion of reason. As a poststructuralist feminist who is also grounded in neo-Marxism, she wrestles with the contradiction of reason as a totalizing system and of reason as a way to liberate people from oppression. More specifically, she critiques ways that reason has marginalized and excluded based on ethnicity or gender, for example, even while she has also called on reason for its powers of emancipation (p. 104). And drawing on Ingram (1990), she cautions that the dialogic process about which Habermas speaks, assumes that all

> involved in such communication are committed to 'truth, rightness, and truthfulness (or authenticity)—whenever they try to reach a mutual understanding...and that rational persons are inherently oriented toward something like an unconstrained democratic community.' (as cited in Kohli, 1995, pp. 104–105)

She warns that "communicative reason" must take into account "the *persistent social differences* that result from systemic practices that create unequal and difficult conditions for people to flourish...[i.e.] the social effects of oppression" (p. 108).

In upacking oppression, Kohli would reconstruct an "understanding of communicative rationality" to include a theory of oppression to raise awareness of how many begin with an uneven playing field and don't have the "dispositions" for engaging in "rational discourse"[2] and "why some have been 'defined out' of the entire process" (p. 108). As well, she expands on the notion of "communicative rationality" with the reminder that those who wish to further democratic participation within a

> 'rational' society need to understand the pervasive and persistent existence of internalized oppression and how it shapes virtually every social situation, every dialogue, every communicative interaction. (p. 109)

Individuals who live with being different—"women, working class people, people of color, or gay and lesbian people, 'internalize'" what is spoken and written about them so that they develop "negative self-images," which translate into ways of acting and being in the world that can place them at a disadvantage in certain types of dialogic interactions. "[P]owerlessness, anger, fear, or hopelessness" can color their encounters in such a way as to render them ineffective, as well as appearing deficient to people of power or status (p. 109). She suggests a strategy for communicative dialogue wherein people are encouraged to express their feelings

> associated with the forms of oppression they suffer due to their particular location in society. Once the feelings are expressed, listened to, and reflected upon, clearer thinking may result, leading to better communication. (p. 111)

She warns that we must acknowledge "the complexity of the affective dimensions of communicative reason" because "it will no longer do to render these feelings invisible or to see them simply as moral deficiencies" (p. 111).

A "substantive conception of rationality" or "reasonableness" should be an aim for education, according to Burbules (1991, p. 218). He charges the "absolutistic, formal, scientific conception of reason [with] exclud[ing] or devalu[ing] legitimate alternative ways of thinking and feeling" (p. 216). He sees education as fostering certain "virtues" of "reasonableness" through educating people to their membership within community so that a student cultivates a "tolerance for alternative points of view, open-mindedness [and] a willingness to admit that [s/he] is mistaken" (p. 219). He says that educating to engender these qualities would be justified because of their "ability to promote certain kinds of communicative relations" (p. 219).

Bernstein (1991) pushes the notion of pluralism as an alternative way of knowing, and points to the incommensurability between our lives, our disciplines, our "language games" (Lyotard, 1979). He describes an "engaged fallibilistic pluralism" that takes "our own fallibility seriously" so that even though we have our own commitments, we are also "willing to listen to others without denying or suppressing the otherness of the other...or think that we can always easily translate what is alien into our own entrenched vocabularies" (p. 336).

The pluralistic nature of school populations makes it vitally important to foster an open acceptance of difference. We must provide students with the "words, concepts, and theory frameworks" (Bowers, 1993, p. 62) that will allow them to think and communicate within contexts of "reasonable" communities. With that, we must re-think school practices that narrowly focus education on the values of independent thinking and competition in a global marketplace (Orr, 1994). Alternative ways of knowing should be incorporated, also, such as interdependence, openness, flexibility, and tolerance of diversity, so that young people will learn to "situate themselves as members of a language community…and also…[come to] understand their interdependence within the larger biotic community" (p. 63). And throughout, a return to values will be necessary because within any situation there will inevitably arise ethical choices and decisions of responsibility.

Critiques of Cohesion

The "impulse" toward a unitary story is a critique leveled against ecofeminist and deep ecology discourses extolling a cosmological view (Quinby, 1990). Even though coming out of different traditions, the views of many poststructuralists[3] and feminists overlap in distrusting the idea of a cosmology based on an inherent universal connection in a meaningful and ordered way. Contingency and particularity are common themes throughout poststructuralist and feminist discourses ruling out a reliance on cohesion. They question whether the idea of a cosmology is not just presenting a new metanarrative to replace the modernist "story"—a Eurocentric, rationalist view with its roots in Enlightenment Europe. Metanarratives that attempt to predefine who human is in terms of a discourse of universals, cohesive, unitary doctrines, or essentializing metaphysics are considered as naive or arrogant (Bloom & Munro, 1996; Sawicki, 1991; Walkerdine, 1990).

It is with good reason that some question the idea that the earth is a "living organism," possessing "organic unity." Holism has led to totalizing systems that have justified abuse of particular groups or individuals, rationalizing that they were acting in the best interest of "the whole." Nazi Germany's totalitarian genocide was carried out under just such a guise. To varying degrees, a similar rationale has been misused to subjugate and control those in subordinate positions throughout history. Domination has been "justified" as a "natural" occurrence, mirroring hierarchical models evident in the natural world, a rationale for control further reinforced by Darwin's "legitimating" perspective of "survival of the fittest." The dark "underside" of this apparently innocuous view on the "unity" of the earth's relations, demonstrates how any totalizing discourse can become dangerous when held naively as unproblematic and thus beyond doubt. (Foucault, 1978; Horkheimer & Adorno, 1972).

Poststructuralists' disavowal of a fixed or determinate framework (Foucault, 1978; Quinby, 1990) and feminists' attention to the particularities of contextualized embeddedness (Munro, 1996) work to undergird one another's positions with points of intersection. Naming ecofeminism as a site of resistance against hegemonic discourses, Quinby (1990) warns against

> essentialist tendencies within ecofeminism [to] speak of a monovocal subject, Woman; of a pure essence, Femininity; of a fixed place, Nature; of a deterministic system, Holism; and of a static materiality, Body. (p. 126)

She uses the work of poststructuralist Michel Foucault (1978) to argue against "calls for coherence" on the basis that

> resistance movements which become orthodoxy are complicitous with the tendency of power to totalize, to demand consensus, to authorize certain alliances and to exclude others—in short, to limit political creativity. (p. 123)

Quinby (1990) sees ecofeminism as "sites of struggle" in that power is dispersed and circulating through culture and there is no one "source" from which power emanates; likewise "there is a plurality of resistance...in a multiplicity of places" (pp. 123–124). She differs with some ecofeminists' desires for unification (Spretnak, 1993 and 1997) in that totalizing theories and centralized practices tend to make "social movements irrelevant...vulnerable...[or] participatory with forces of domination" (p. 123).

Ethical considerations are especially important when defining a community as an open system. Within living systems, according to Berry (1988), "every expansive life force should have arrayed against it limiting forces that would prevent any single force or combination of forces from suffocating the other members of the life community" (p. 116). There has to exist a balance between play and responsibility. But how to strike such a balance? What does the basic premise of "self-organization," implied in the definition of "living systems," mean when applied on a societal level to human beings? Is the "self-regulation" that is said to balance and equilibrate living systems also translatable into acts of racial "cleansing" or to the annihilation of six million Jews by Nazi Germany in World War II? Do we trust the self-regulation of a "system" to replace ethics? And who is responsible for answering questions of this nature? If decisions remain with the "dominator" mode of consciousness, there is little hope that we will have more than "business as usual."

An "open system" as a model to conceive of more relationally grounded social practices will necessarily carry, also, parameters of right or wrong, times where we must step in and intervene. A system can't be totally open because we are dealing with human lives. When decisions for responsible action arise, actions will come down to a question of values, so we must begin by delineating what is responsible within an ecological frame. While beyond the scope of this book, foundational to beginning such a process of deliberation is the understanding that to engage in ecospiritual praxis, one takes into account that humans, their society, and the planet are all mutually interdependent. As such, an injustice to one is an injustice permeating the integrity of the larger relationship, the planetary body. And it recognizes that acts of injustice are not only committed against "Others" but also harm the one initiating the negative action.

Other questions for consideration include the following: When we place people/cultures/societies onto the template of "natural order," what does that mean for legislation, for justice, for equality? Will there be those who are left out? Who are pushed aside? Who are not "naturally selected?" We have seen how it is possible for the "free hand of the marketplace" to exclude and dispose of human life without conscience. Where does human agency come in—to counter misogyny, homophobia, racial prejudice, and the accompanying violence committed in the name of eliminating difference? Was not the idea of unification all too often used to rationalize the colonialist appropriation of life, liberty, and resource from "third world" peoples lacking strength and number to resist? Does the argument for unification flatten difference to the extent that the richness that flavors diverse cultures be dissolved into the mix, or worse, erased through acts of violence?[4] And are we to assume that as we move into "being" that some "natural" ethic will emerge?

There are indeed no easy answers to these and countless questions raised by the consideration of a holistically informed, ecological framework for living. Fears of metanarratives and totalizing discourses are understandable in light of their use historically to exclude and to marginalize. The disruption of hegemonic foundations and totalizing discourses such as patriarchy is paramount to re-creating the present toward the possibility of living in long-term ecological sustainability with life on the planet.

Working for justice is part of the cosmological process (Fox, 1995) and part of intentional living, or mindfulness, is an awareness of the historical context within which any moral dilemma occurs. Promise lies, I feel, in the capabilities of many ecofeminist theorists (Macy, 1991a; Merchant, 1996; Salleh, 1997; Spretnak, 1997) whose work disrupts within the spirit of *hope*, de-sedimenting, challenging assumptions with the proactive intention[5] of reframing and reworking the structure. Stephanie Kaza (1993) says that an awareness of the contextual ethics involved in each unique circumstance within a web of relations connotes "a shift from emphasis on rights, rules, and principles to a conception of ethics grounded in specific relationships" (Kaza, 1993, p. 61). Life is determined at the nexus of relations, an arising from the bottom up, the site at which ethical choices are possible. The agency necessary to initiate change, then, will most logically come in the form of local struggles as exemplified by Joanna Macy's (1991a) concept of "the greening of the self" (p. 183). Macy defines the self as a

> metaphoric construct of identity and agency, the hypothetical piece of turf on which we construct our strategies for survival, the notion around which we focus our instinct for self-preservation, our needs for self-approval, and the boundaries of our self-interest. (p. 183)

Macy sees our perception of the self as making a "shift" to "wider constructs of identity and self-interest," to what she calls the "ecological self...co-extensive with other beings and the life of our planet" (p. 183).

For example, human movements to stop deforestation, or to intervene to stop the slaughter of marine mammals, employ strategies that often risk the lives of activists involved. Part of the momentum of these and countless instances of agency is the activists' extension of the "sense of self to encompass the self of the tree and of the whale" (p. 184). In moving to connect, to extend relationality and bridge the dualistic thinking that distances "tree and whale" as objects to be exploited, they are "no longer removed, separate, disposable…pertaining to a world 'out there'" (p. 184). The greening of the self involves a moving beyond the "separateness, alienation, and fragmentation" constitutive of the dichotomous self and other. It calls for a renewed "sense of the sacred" as the generation of a "profound interconnectedness with all life" (Macy, 1991b, p. 184) and offers a cosmological view of "an encompassing self, that deep identity with the wider reaches of life…as motivation for action" (pp. 184–185). If one understands God as the "dynamic process by which life pours forth in all its variety of expressions," Rosemary Radford Ruether (1996) suggests that

> [e]vil also exists in these relations, not as something willed by the creator…but as a way humans interrupt this life process by seeking to control it, to lay hold of its power and wealth for the few against the many. The struggle for liberation,…[then] is the struggle to overcome…distorted relations, to renew human life in its context of relations among people…and the earth community. (p. 10)

For these and other reasons, I do not see a cosmological view as necessarily counter to emancipatory practices of liberation strategies (Macy, 1991a), since it distrusts metanarratives that tend to be closed and determined structures. A processional view of "dependent co-arising" provides some boundaries that lend integrity to the system and make it a whole, even while the openness across boundaries allows systems to overlap. With no finality, no absolute determinacy, no archimedean point, a relationally based systems-view is as perpetually unfinished as the living of life itself. As an epistemological matrix for our knowing it offers *relative* certainties contingent on the particular, the context-bound, sites where decisions of responsibility are possible—grounded within the context of an ecospiritual ethic rooted in interdependence, justice, and ecological sustainability.

Notes

---

1. Although beyond the scope of this project, holographics is a fascinating area of research which is emerging along these lines (see Bohm, 1983 and 1985; also Talbot, 1991).

2. For further discussion see Siegel, 1991.

3. Poststructuralism is a branch of postmodern philosophy that emerged on the intellectual scene in Paris of the 1960s as both an "assault on structuralism and also an outgrowth of it" (Pinar et al., 1995, p. 452). Poststructuralist scholars, for example Jacques Derrida, Jean-Francois Lyotard, and Michel Foucault, are often linked with the French school of Continental philosophy, which comes from a tradition of discourse analysis and literary criticism (Sturrock, 1986) and all write widely within the social sciences. Hoy (1988) says that poststructuralists such as Derrida and Foucault "aspire to break with modernity" by breaking it down and "showing its self-delusions" (p. 20).

4. Racial "cleansing" is a horrifying example that has surfaced in recent years.

5. By proactive, I mean "acting in anticipation of future problems, needs, or changes" (*American Heritage Dictionary*, 1996, p. 1443).

Chapter Five

How can you buy the sky? How can you own the rain and wind?...We are part of the earth and it is part of us...Teach your children...[t]he earth is your mother. What befalls the earth befalls all the sons and daughters of the earth...We did not weave the web of life, we are merely a strand in it. Whatever we do to the web, we do to ourselves.

—Chief Seattle, 1850s

All life forms have an intrinsic worth and a right to evolve freely on their own terms. Humankind is one among millions of other species. It does not have the right to push other species to extinction or to manipulate them for greed, profit and power without concern for their wellbeing.

—Vandana Shiva, 2000

Place, Community, Cosmology: At Home on the Planet

There's a damp-moss, humid-earth feel this July morning, not unique to Southern Louisiana. It penetrates the skin and leaves a residual stickiness, telling of the day's impending heat. Light filters onto my page through sapling limbs on the ridge above, back toward the east and the rising sun. Across Bayou Fountain I notice a large, dark form perched high on a grapevine that spans the distance between two trees. I bring up the binoculars to frame a large screech owl into view. Rust-colored spots mark a full, white breast. A reddish cowl of feathers meets a darker mask, rises up into two tufted-horn points, and extends the full length down his back. He appears to be watching me with deep round eyes, attentive. After some time he turns slightly on the branch, ruffling his feathers. With a sudden movement belying his size, he spreads into flight and sails down and along the shadowy channel of water. I suspect that it is time for him to sleep. What is left from this experience to color my being? How am I a part of the breathing-out-and-breathing-in of the cycles of life that surround me? What does my relation with this land offer that nothing else can come near? I *am* this land that I inhabit.

Do you consider yourself an "indigenous" person? William McDonough (2001) asks this question, and it brings me to think of the relations marking me as who I am, as indigenous to this place as any creature threading the web of the ecological balance. *Indigenous* or *native* refers to one who belongs to or is "connected with a specific place ...by virtue of birth or origin...intrinsic (*American Heritage Dictionary*, 1996, p. 1203). When does one come to thinking of the self as indigenous to this planet, because "the minute you start to think of yourself and your children as native to a place, your sense of...responsibility start[s] to shift" (p. 1). You realize that decisions you make will affect you and those who come after you, for "seven generations" as Iroquois wisdom attests (p. 3). McDonough is a world-class designer whose community and corporate designs recently won him the Presidential Award for Sustainable Development. If one considers the meaning of "design," we are thinking in terms of *intentions*. Design is intentionality. Design is thinking about what it is that we want to create and from there making a conscious choice to carry that creation to fruition. McDonough calls design "strategic intentions" (p. 2). And design typically brings with it a choice.

I'd like to touch on a story exploring the idea of strategic intentions and also instances where such conscious choices are lacking; remembering that *not* choosing is also a choice. *Ishmael* is Daniel Quinn's (1992) erudite gorilla who recounts a tale of two cultural archetypes, each holding contradictory perspectives on living: one he calls a "Taker" perspective, the other, "Leaver," for their somewhat neutral connotations. Yet, they reflect more loaded cultural terms—*Takers* to mean those cultures that Westerners often label as "civilized" and *Leavers* those considered by many to be "primitive." Takers see the earth as a resource for their own use, operating from a principle that "the world was made for man [sic] and man was made to conquer and rule it" (p. 72). Taker cultures exercise a "power-over" means for taking or doing those things they can *rationalize* as correct from a view that assumes theirs to be the "right" way to live. This position is thought to give them the right, also, to judge how others should live. A Leaver culture is more in keeping with a relational point of view, recogniz- ing the rights that other life forms have to survive, taking what is needed and leaving the rest. Leavers are less inclined to believe one *can* own property, but would rather live *with* the earth, than to possess it. The words of Chief Seattle above put it elegantly. A Leaver view is commonly attributed to beliefs within many tribal value-systems, a perspective sometimes termed "indigenous mind" (LaDuke, 1996).

The story's protagonist, Ishmael, able to communicate his wisdom telepathically, tells the story of an aviator who has not considered gravity in designing his aircraft. The pilot, zealously sets off in a plane that is not in compliance with the laws that make flight possible in the first place—it cannot fly—yet he doesn't know that the ground is fast "rushing up to meet" him (Quinn, 1992, p.108). Ishmael draws an analogy here to human Taker cultures who, charmed with the notion of flying, take off in their "Taker Thunderbolt" without working to consciously understand the way the system operates in the first place, the biological balance of life. Without knowing it, they are nose-diving in free-fall because their Taker craft is simply not in compliance with the laws of the biological community, earth, "and that fall is about to end" (p. 109).

Difference as Bridge to Relationality

Ecospiritual praxis requires a shift from an ego-centered perspective to an eco-centered one toward more integrative ways of understanding and experiencing human/earth relations. In exploring the word *relate*, synonyms are "join, combine, unite, link, connect,...[and] associate" (*American Heritage Dictionary*, 1996, p. 971) suggesting, also, that a vital component within relationship must be the presence of *difference*. Without difference, what would we have to "combine?" What would there be to "join?" Difference is inherent in the notion of relations. And within this earth relationship, biologist Vandana Shiva (2000) reminds us that "[b]iological diversity and cultural diversity are intimately related and interdependent" (p. 8). Diversity is crucial to maintaining life on this planet, yet, "according to the dominant paradigm of production, diversity goes against productivity," (p. 16) whether you speak in terms of human cultures or varieties of plant or animal species. The industrial/technical model dressed up as "globalization"[1] continues to create an imperative for uniformity and monocultures" (p. 16). Proponents of globalized markets often present themselves as "guardians of the 'world community' or of universal human rights and the free world market...while, claiming the right to exploit local ecology, communities, [and] cultures" (Mies and Shiva, 1993, pp. 8–9). Agribusiness and the biotechnology industry places relative worth of species on their material use as resources for production and consumption. Yet diversity is what keeps this planet alive, making such a view narrowly shortsighted. It is alarming, then, that the extinction rate of species now stands at "27,000 per year—1,000 times the natural rate—and [that] human greed and desire for profit are the primary causes of most of these extinctions" (p. 11). According to Shiva (2000), "biodiversity is the indicator of sustainability" (p. 6). She points to "species diversity" as the "richness of an ecosystem,...[and] ecological interactions between diverse species the key measure for ecosystem diversity" (p. 12). Within the global paradigm, diversity existing at the level of local communities is also at risk. For example, Mies and Shiva (1993) say that

> local cultures are deemed to have 'value' only when they have been fragmented and these fragments transformed into saleable goods for a world market...food becomes 'ethnic food', music 'ethnic music', traditional tales 'folklore'...thus [are cultures] dissected and ...commodified...thereby procuring a standardization and homogenization of all cultural diversity. (p. 12)

I return to the value of difference as bridge to relationality. We stand at a crossroads facing an imminent question: How will we live with the earth in this new millennium? (Hogan, 1996). Although the rhetoric of globalization conjures up images of one-world-one-people equaling a better life for all, we must consider the underlying "design" in determining best practices. What are our strategic intentions? Do we consciously consider the choices we make within an ethical context that is larger than our own self-interest? If actions are motivated by "reductionist knowledge, mechanistic technologies and the commodification of resources" (Shiva,

2000, p. 8), it is possible that we walk a path based on profit and greed. It is possible that our basic design is rooted in the Cartesian "severance [of] human knowing and being from any sense of earthly embodiment, obligation, necessity, or ecological consequence" (Jardine, 2000, p. 89). McDonough (2001) takes us back to the "first industrial revolution as a design assignment":

> I'd like you to be involved in a system that produces billions of pounds of highly toxic hazardous material...and puts it into your soil, air and water every year... Measure prosperity by how much natural capital you can cut down, dig up, bury, burn or otherwise destroy. Measure productivity by how few people are working. Require thousands of complex regulations to keep you from killing each other too quickly. [And] destroy cultural and biological diversity at every turn seeking one-size-fits-all solutions...Can you do that for me? (pp. 10–11)

I agree that "it is time for a new design" (p. 7). We have lost sight of our intrinsic connection to the planet. We're piloting in an aircraft that is not in compliance with the laws that make flight possible in the first place—and we're heading for a fall.

Ecospirituality as Relational Knowing

The writing of this book was an unfolding. I did not sit down and say, I'm going to write about *relationality*. And yet the topic would come up whenever I would write. I'd think, "Well, so what? Why is this so important to me?" I continued for months creating in the midst of this nagging question: "What *about* relationality?" And then there came a turning point, it happened in a dream.

I went to sleep one evening, restless from the question I was living. I awoke in the night with it tossing and turning in my head, my whole being infused with a fear that often comes from the unknown. And so with great emotion, *I asked inside myself,* "Please give me some understanding that'll tie this all together, and give purpose and meaning for my work." On falling back asleep, I experienced a dream that was vivid and powerful, vital in its impact. It was one of those "big dreams" that leaves an impression upon awakening. As my eyes opened, there remained the trace and some vestige of a vision etched on my awareness. The message was clear: a "Holy Mass" were the words left reverberating through me. I reached over to the night stand and jotted this description:

I was not myself alone, but I was also many, and we were all a part of this large mass *upon the planet doing work, sowing seeds of creativity and life, each in our own way. A story I had heard long ago comes to mind, that of Indra's Web with its image of a jeweled skirt reaching over the planet. The story applies here, for in the dream I saw clearly that each one of us—each small group of us—was connected with all of the others, and we were illuminating the planet with our light. And by affecting any one of the jewels in the web, all of them were affected, somehow, all connected by our consciousness, a relational way of knowing.* And on awakening from this dream, I knew beyond any question that "relationality" would be the way of knowing I'd work to foster in my life.

The sense of connected consciousness I describe was an awareness that "god" was *in the relations*. It was an integrative knowing, an activation of relations within all of our capacities to know. It was not a narrow view of knowing—one that science might parcel out as valid. Rather, it was a knowing that integrated the entire being and extended between and among all other beings. The words and image of a "Holy Mass" was the impression I was left with on awakening. And from that point on, the conviction of my spirit was deepened and further brought to bear upon this work.

The word *ecology* is based on an awareness of relationality—as a science, "ecology" deals within the realm of relations. The term names the connection between human beings, their communities, and the cosmos. From the Greek, *oikos,* ecology means "home" or "dwelling." Jardine (2000) notes that, "ecology concerns…what is properly responsive to the place in which we find ourselves…It requires action and thought which preserve the integrity of the place that houses us" (p. 30). The place wherein I dwell grounds me as who I am: my body and all of its varied dimensions. Ecology names the relations I sustain within a social community with their ties to place and cultural context. And it frames my connections within an ecologically self-maintaining system that is open and dynamically generative, the larger planetary body of my origin which sustains each of its parts within an emerging, creating, cosmology of potentials.

At this time in earth history, Berry (1988) says, there is a mounting energy toward a new way of looking at life as interrelational, an emerging ecological sensibility. This momentum is reinforced in part by a resurgence of interest in varied forms of ancient spiritual teachings, including those of particular indigenous peoples characterized as being "more likely to revere interdependence, cooperation, and reciprocity than western cultures" (Jacobs & Jacobs-Spencer, 2001, p. viii). Also, through the increased understanding of the workings of the universe coming from postmodern science, Berry (1988) notes that a "functional cosmology," or creation story, is giving these interrelationships new meaning. And within this meaning is

> the deepest mystery of the universe: the revelation of the divine. To deepen this experience of the divine is one of the purposes of all spiritual discipline and of all spiritual experience. This sense of communion at the heart of reality is the central force bringing the ecological age into existence. Thus the birth of a new overwhelming spiritual experience [is occurring] at this moment of earth history. (p. 121)

In my estimation, spirituality is the ground from which the fire of the creative spark of life emanates. It is the unnameable cohesive property that connects all things and brings to life the multiplicity of forms between the microcosm and macrocosm. It constitutes a process of creation and dissolution occurring at all levels of life that manifests within human beings as a continuing journey toward that "becoming" that is never finished. I think that such a view of spirituality is ecological in a broad sense in that it recognizes all forms of life on the earth as sacred and respects them as such (Mies & Shiva, 1993). An ecospiritual perspective acknowledges the human relation within the environment as mutually sustaining and honors an organic, deeply felt spiritual connection as the context linking the individual within place, community, and cosmos.

I *am* this place I inhabit. I live each day in relation with the landscape that surrounds me, knowing "that the *deepest* sources of personal and cultural identity are the ecological and geological landscapes that shape and sustain" me (Cheney, 1999, p. 156). And yet, I am *looking for* this relation. I know that it is here and I open my eyes so that I may see it. I carry this conscious intention because I know that it fulfills me; such relations have fed me before. I also see the desecrated landscapes of our "civilized" world, the "landscapes of human culture and humanly transformed nature—broken landscapes that mirror our own brokenness (p. 156), and this I mourn. The importance of place is an ecospiritual awareness, a way of knowing or viewing the world that honors the importance of "past practices, folkways, and traditions...in the creation of new knowledge" (Orr, 1992, p. 31). Pinar et al., (1995) have said that "place embodies the social and the particular" (p. 291). Indigenous peoples of Australia whose knowing is tied to the land, "sing the earth back into existence" (Abram, 1996). They capture the experiences of their lives through an oral retelling, situating their story in place as they move, relating the events to the landscape they pass as they speak. Each place they describe has significance, power, and is always acknowledged at the outset of the telling for the power that place can lend to the tale. Theirs is an embodied knowing of place. Place connotes our situatedness in geography, community, and within a larger cosmology of connections.

Ecospiritual praxis embodies a recognition that the ecological world is an abode for all earthly beings and communities-of-beings of every mineral, genus, and species—all constituting an implicate order, which is chaotic and also integrative in its make up. When viewed from an open-systems perspective, the idea of "community," also, is a multifarious construct. It is not only integrative, not only sharing common ground; the image I drew in chapter 4 was open, multilayered, multidimensional. The communities in which we dwell are not closed systems. They are overlapping worlds-within-worlds, with infinite layers of cohesion and disassociation, intersection and disjunction. They combine and collide into multiple forms. Derived from a Latin word *communis*, or "common," the term "community" has a broad range of varied definitions. The term can be applied to a group of people living in the same location and under the same laws of govern-ment. It can be said of people with common interests, such as a scientific community; to those gathered in a communal form of fellowship or sharing, as to a community of worshipers; or it can be used to designate an ecological region populated by particular species of animals or plants. "Bioregionalists" recognize community demarcations in accordance with how the land, itself, presents them. Rather than arbitrary human-imposed boundaries, they locate communities within parameters of watersheds, plant communities, local places viewed "ecologically rather than, as usual, socially and politically" (Weston, 1999, p. 195). Community and landscape, here, are inextricably intertwined.

Communities span lifetimes and locations, crossing historical borders of time and place to include those who have shared some common understanding. Communities cross chasms of contrast and difference between and among those who speak for common causes, even though they come out of different backgrounds, traditions, or historical contexts. Each community is as unique as the populations composing it. Some can be said to have a "special" quality that gives them strength, keeps them vibrant and healthy. In ancient times, at the center of every human community was a circle of fire that was always kept burning. It was the place where people gathered to find warmth and sustenance, a communion of family and friends, a place to discuss the events of the day, to debate, argue, take refuge from enemies. Over time, the ring gave way to the hearth, which from the Old English, *ker,* means "heat," "fire," "ember." The hearth is defined as "the floor of a fireplace;" and also connotes "family life" and "the home" (*American Heritage Dictionary*, 1996, p. 834). The hearth is the floor on which the fire is laid, it is the ground that contains the heat and returns its warmth. And I would suggest that what strong and vital communities may have in common is that sense of the hearth—the context within which that fire is laid and which also holds the heat and returns that warmth which vitalizes from within. The core of this crucible, I would say, emanates from an awareness of the deep connection we share within an interdependent matrix that is larger than ourselves.

The basis for a "living spirituality," Fox (1995) says, "is an emerging cosmology that can be seen within the shift "from the modern to the postmodern era of spirituality" (pp. 7–8). There is evidence of this shift within the widespread disruption of modernism's foundational maxims such as Newton's mechanical universe, "Descartes' dualism[s]" and "Bacon's dualistic ferocity against nature and women" (p. 8). A shift to a cosmological view is based on a relational way of knowing that recognizes that

> human beings, the earth and the whole community of life on earth, and finally the entire cosmos exist and are sustained by one breath of Life, one matrix of life-giving relationality in which we live and move and have our being. (Acts 17:28; as cited in Ruether, 1996, p. 10)

It is this relational way of knowing that Fox (1995) incorporates into a fourfold path for viewing humans' spiritual journey, part of the cosmological process in which all living things participate. This fourfold path is not intended as a methodology, but rather as a way to organize thinking around the personal process of spiritual growth. The first of the four paths is an awakening to the awe and wonder that is potentially present in every moment. He terms it the Via Positiva, or what Rabbi Heschel calls "radical amazement" (as cited in Fox, 1995, p. 20), which is to experience the "delight, awe, and wonder...available to all of us on a daily basis...be they in nature, in our work, in relationship, in silence, in art, in lovemaking, even in times of suffering" (p. 20). The second, the Via Negativa, is a letting-go of the need to always control, a willingness to experience the mystery, the chaos, and could include practices in which we "let go of sensory input" such as meditation or fasting. Fox says that letting go of words and images is important "to a postmodern spiritual practice because so much of the modern era is wordy...[perhaps] for the fear in a patriarchal and 'enlightenment' era, of the dark, of silence, of what cannot be controlled" (p. 20). With the Via Negativa is also the recognition that "letting pain be pain is an essential ingredient of learning from pain and in experiencing the dark. It is also essential for letting go of pain" (p. 21). From the Via Positiva and the Via Negativa comes the third path, the Via Creativa—creativity itself—the central spiritual principle in a living cosmology that acknowledges our place as part of creation, as well as creators. The third path is vital in a postmodern era for carrying us "beyond the notion that the universe is completed or is a machine in motion" (p. 20). The fourth, the Via Transformativa, flows from the creative principle and, according to Fox, is rooted in the issue of *compassion*, as a response to an interdependent universe. He says that compassion "means both celebration and healing by way of justice making" so that we acknowledge our capacity, indeed our responsibility, to "interfere with the causes of injustice" (pp. 20–23) and with the ways separation fuels patriarchal power relations such as sexism, racism, individualism, reasonism, scientism, and naturism. Fox explains that "while creativity lies at the heart of the universe and at the heart of the human psyche and spiritual journey, it finds its fullest expression in the transformation of society itself" (p. 23).

Embracing a cosmological view can include a willingness to "let go," and open to the life-giving power of the sacred within. Not a distant hierarchal ruler rather it epitomizes an "underlying font of being" (Ruether, 1996, p. 10) that

> upholds the life-process of all creation as it...continually wells up and is renewed. This life-process unfolds through a dynamic of diversification, interrelation, and communion...shap[ing] everything that is—[from] the cosmos with its many stars and galaxies, to the rich variety of plants and animals on the earth, to the diversity of human cultures, to the interactions of two people with each other, and finally, to one's relationship to oneself as embodied spirit. (p. 10)

A Praxis of Relationality and Difference

A move toward relational ways of knowing rests upon the valuing of difference as bridge to more relational points of view. An awareness of our cosmological embeddedness within an infinitely creative universal order, Lydon (1995) maintains, is predicated on complexity constituted by subjectivity, difference, and interrelation (p. 78). Who we are and how we relate is marked by a notion of "difference," therefore, relationality as a frame for an ecospiritual educational praxis brings with it a need to address the issue of "difference." This view will necessitate a shift in the ways that "Others" are perceived—away from the ranking model of patriarchal hierarchies, toward ways of knowing that perceive *difference* in more egalitarian terms. It also must acknowledge the intrinsic value to be found in all "Others." The evidence is alarming that "environmental degradation stemming from the exponential growth in resource consumption and human population...pose[s] very real threats to the earth's biological support systems" (Eckersley, 1992, p. 12). Personal and social alienation continue to pervade our world in forms such as "decaying inner cities, insensate violence, various addictions, rising public debt, and the destruction of nature all around us" (Orr, 1994, p. 51). I believe, with Don Jacobs (1998), that the basis for many of the destructive elements evident in Western culture is an "estrang[ment] from the wisdom and spirit that connect us to the natural world" (p. 229).

Education has a vital role to play in reversing the trend from one of destruction to one of renewal. With David Orr, I propose that the root of these imminent problems lies in the way we think, calling for a re-vitalization of the "institutions that purport to shape and refine the capacity to think" (p. 2). Many of the world's crises begin with an education that

> alienates us from life in the name of human domination, fragments instead of unifies, overemphasizes success and careers, separates feeling from intellect and the practical from the theoretical, and unleashes on the world, minds that are ignorant of their own ignorance. (p. 17)

In order to vitalize Western ways of knowing it will first be necessary to re-evaluate entrenched cultural beliefs and practices in order to discern those which have proven personally, socially, and ecologically unsound. Only when we've exposed our knowing for its problematic assumptions, will we be able to move toward beliefs and practices that support more holistic and ecologically sustainable ways of life. Allow me, then, to explore some alternative pathways that relational ways of knowing might offer curriculum and teaching.

Alternative ways of viewing "Others" might allow us to negotiate gaps wherein differences divide man from woman, white from black, culture from nature (Aoki, 1993). Differences set us apart. Teachers and students come-into-being between and among their differences. Differences signify a self and an other, which may bring up a wall dividing being from being, inhibiting the relationality that is the fabric of life. Ted Aoki draws on

Deleuze's (1977/1987) comparison between *difference in degree* and *difference in kind*. For Deleuze, the former is a way of seeing "in terms of more or less": more power, more money, more beauty, more this or that, for example, where one is *more* of whatever making the other *less* of the same. Understanding difference as more or less is a competitive view which is usually couched in hierarchical judgments. And, says Deleuze, "each time we think in terms of more or less, we have already disregarded differences in kind between the two orders, or between beings...(pp. 20–21), where comparison is irrelevant. Each is uniquely who and what they are, each bringing richness into the classroom mix. To see difference in terms of degree (more or less) where "there are differences in kind is perhaps the most general error of thought, the error common to science and metaphysics" (p. 21) and also education.

Learning to navigate within ever-shifting realms of difference is among the most important work a teacher will ever do. Aoki (1993) reminds us that the experience of the teacher is one of in-betweens, dwelling as he or she does, in a world textured in multiplicity, a classroom of unique entities whose lives are brought to bear in and among the differences of "others." A teacher walks the space between curriculum-as-plan and curriculum-as-lived-experience, the latter being a "poetic, phenomenological, and hermeneutic discourse in which life is embodied in the very stories and languages people speak and live" (p. 261). Generally, curriculum-as-plan comes from outside of the classroom created for a "homogenous" realm and "faceless people" (Aoki, 1993, p. 261). It is "imbued with the planners' orientations to the world...[their] interests and assumptions about ways of knowing" and understanding teachers and students (p. 258). This instrumental curriculum frames a "set of statements...in the language of goals, aims, and objectives...ends, and means" (p. 258). The teacher moves in the margins between "plan" and "lived curriculum," between multiple entities expressing their differences.

The world of curriculum-as-plan is different in kind from that of the curriculum-as-lived experience. In his/her wisdom, the teacher knows that there are many lived curricula, as many as there are self-and-students, and possibly more. And within the differences, the world of lived curriculum is a world of multiplicity. In exploring this world, we could posit "identities in the landscape…a habit of modernism grounded in the metaphysics of presence . . .a view that any identity is a pre-existent presence…we can re-present by careful scrutiny and copy" (p. 260).

In a postmodern turn, Aoki "reconsiders the privileging of 'identity as presence'" to instead understand teacher/student identities as the "ongoing effects of our becomings in difference" (p. 260). This notion of "identity as effect" displaces "identity as presence" in a movement that "consider[s] identity not so much as some*thing* already present; but rather as production, in the throes of being constituted as we live in places of difference" (p. 260). This place wherein the teacher dwells is a world of multiplicity and between-spaces, "betweens" that are sites of difference. Aoki suggests that a "reattunement" of perspective on identity from "identity as presence'" to "identity as effect" moves multiplicity from being viewed as "multiple identities" to a more processional coming-into-being "as we live in places of difference" (p. 260).

It is these places of difference in the classroom between plan and life, between self and other, the margin spaces between differences in kind, which texture the curricular landscape with color, richness, and diversity. "In a multiplicity," as we have said, "what counts are not…the elements, but what there is between," echoing Deleuze (1977/1987), every multiplicity grows in the middle. . ." (p. viii). He draws on Heidegger (1981), who says that "the relationship between teacher and taught" forms in this embrace, ideally where meaning is a negotiation, a generative, dialogical making-into-being suggestive of difference in kind, "where there is never a place…for the authority of the know-it-all or the authoritative sway of the official" (pp. 15–16; as cited in Aoki, 1993, p. 266). To reorient our understanding of difference from difference in degree to difference in kind is to "embrace the otherness of others" (Aoki, 1993, p. 266).

Difference as a bridge to relationality is evident in Murray Bookchin's (1982) exploration of research documenting certain preliterate communities. He calls them "organic societies because of their intense solidarity internally and with the natural world" (p. 44). Evidence reveals that in these cultures, there is a way of viewing the world in which "people, things, and relations " are seen "in terms of their uniqueness rather than their "superiority" or "inferiority" (p. 44). Communities with this sort of perspective on difference are more likely to view "individuals and things…[as] not necessarily better or worse than each other…[but] simply dissimilar…[each being] prized for…its *unique* traits" (p. 44; emphasis his).

The status we accord the Western ideal of "individuality" is absent in this example, meaning that within this perspective there is a lack of the "fictive 'sovereignty'" (p. 44) that mythologizes the notion of the autonomous individual in the twentieth-century West. The world perceived in this so-called *primitive* outlook is "as a composite of many different parts, each indispensable to its unity and harmony" (p. 44) and representative of a part-to-whole relation. In that all depends on the strength of the community for survival, individuality is experienced "more in terms of interdependence than independence" (p. 44). Moreover, in those cultures that still operate out of this perspective, Bookchin says, the linguistic structures for possessive, dominating, and coercive types of behaviors is nonexistent. To illustrate, he draws on the work of anthropologist Dorothy Lee (1959) who examined the syntax of the Wintu Indians [sic] and found that

> a Wintu mother…does not "take" a baby into the shade; she *goes* with it. A chief does not "rule" his people; he *stands* with them. [Instead of saying] 'I have a sister,' or a 'son,' or a 'husband,' …*to live with* is the usual way in which they express what we call possession, and they use this term for everything that they respect, so that a man will be said to live with his bows and arrows. (p. 45)

By using this example, I do not mean to reify premodern cultures as if they were superior to our own. I only suggest that they may have some qualities from which contemporary Western culture might learn in rethinking the idea of difference, not as better or worse—differences in degree—but as differences in kind, valuing "variety…within the larger tapestry of the community—as a priceless ingredient of communal unity" (p. 44).

I propose that we might re-consider "difference," then, as a foundation for "relational" knowing, based in an openness that recognizes that it is through difference that we come to know ourselves and our world as a part-to-whole relation. Bateson (1979) demonstrates that "perception operates only upon difference...[and that] all receipt of information is necessarily the receipt of news of difference" (p. 29). He illustrates the point by taking a piece of chalk and grinding the tip of it into a thick raised spot on a smooth blackboard. If a person were blind, the only way to discern the mark from the board would be in feeling the difference, the roughness of the chalk against the smoothness of the board's surface. It is the difference *that makes knowing possible* (pp. 96–99). And it is difference, diversity, which constitutes the multiplicity of particularities—"the many"—comprised within "the one."

Ecospiritual Praxis and Curriculum

Helping young people come to value difference and interrelations is to acquaint them with the contexts within which they live—their geographical situatedness within a particular area or region. Immersing students in an outdoor setting for a span of time, Orr (1994) says, contextualizes education in the surrounding environment and helps them see that "[n]atural objects have a concrete reality that the abstractions of textbooks and lectures do not and cannot have" (p. 96). I agree with Orr "that nature has something to teach us" (p. 95). "Living" a course in the out-of-doors, along a river for example, provides for an experience wherein the pace of life and learning slows, allowing a sort of "mindfulness" to ensue and to be cultivated, a space conducive for "a deeper kind of knowing to occur" (p. 96).

A recognition of "place" would contextualize curriculum within an awareness of the "balance of life systems, the imaginative flexibility and adaptability of nature, and the integrity and creative harmony of the ecosystems" (Lydon, 1995, p. 77). Orr (1992) suggests a reconceptualization of "the purposes of education in order to promote diversity of thought and a wider understanding of interrelatedness" (p. 129). He says that

> places are laboratories of diversity and complexity, mixing social functions and natural processes.... If the place also includes natural areas, forests, streams, and agricultural lands, the opportunities for environmental learning multiply accordingly. (p. 129)

An ecospiritual basis for grounding curriculum in a sense of place would "center educational inquiry upon the Universe as a whole and humanity as a part of this entirety" (Lydon, 1995, p. 74). Awakening within children a conscious awareness of place implies possibilities for the "awe-filled knowing" of their "place in the universe"—moving curriculum beyond a static, institutionalized "form of knowledge of how the world is" (Bohm, 1983, pp. 3–4; as cited in Lydon, 1995). And it could be used to demonstrate how ties to community are embedded within a context of place, but how living things *also* exist within multiple communities that sometimes overlap, combine, or transcend place in a variety of ways.

The idea of community has been important for education in that, according to Dewey (1897), "school is simply that form of community life in which all those agencies are concentrated that will...[bring] the child to share in the inherited resources of the race and to use his [sic] own powers for social ends" (p. 126). Bringing those powers to bear on the relationships within communities can be an empowering experience of learning for young people. According to Rockefeller (1989), Dewey saw schooling as a microcosm of the larger macrocosm of the culture. Through the socialization process within schooling, Dewey envisioned that students would come to understand the workings of a "community in which all individuals are provided with the opportunity to develop and employ their special abilities" (p. 307). In this way, the "individual...finds realization of the self and the community is sustained" (p. 307).

Communal forces that unify people into groups do not always result in ends deemed positive within mainstream culture. Lisa Delpit (1997) has said that the same type of desire that elicits participation in school and civic clubs, church organizations, and family activities is also what draws young people into street gangs: a desire humans have to belong, to be a part of something larger than the individual self. As social beings, humans are "interconnected with their environment" and have a "basic need to feel they belong to the larger whole" (Rockefeller, 1989, p. 307). In this way, communities intertwine the personal with the social. They provide forums for ingenuity, drive, and innovation, empowering young people to believe in the possibilities that people working together can attain. Conversely, communities can provide for the same sorts of creative drives to be directed toward destructive capacities also. The power of a group joined and focused toward common goals can fall anywhere from creative genius to extremes of destruction. A determining factor behind the direction a community may take is the ways of knowing and of meaning making that underlie its beliefs and practices. These processes are culturally driven, and as such, education has the opportunity to play a vital role in re-shaping the ways in which meaning is made in society.

In *The Culture of Education*, Jerome Bruner (1996) proposes a kind of "institutional anthropology" that would help "formulat[e] alternative policies and practices" for education, taking into account the "reciprocal relation between education and the other major institutional activities of a culture: communication, economics, politics, family life, and so on" (p. 33). With this, he contrasts two theories of intelligence, the "computational model," and the "culturalist" approach. The former is concerned with "information processing" operating within a "rule-bound code." It operates under the premise that "systems are governed by specifiable rules for managing the flow of coded information" (p. 5). While the "computational model" ideally seeks "foreseeable, systematic outcomes," Bruner says that applying its principles to the human mind is difficult in that "knowing is often messier, more fraught with ambiguity than such a view allows" (pp. 1–2). The "culturalist" approach is concerned with the "situatedness of education in the society at large" (p. 33). That means, that in addition to stressing ways education interrelates with other institutions, a culturalist approach considers "crisis problems" of education "like poverty and racism."[2] It investigates the part that "schooling play[s] in coping with or exacerbating the 'predicament of culture'" (p. 33). A culturalist approach "takes its inspiration from the evolutionary fact that mind could not exist save for culture" (p. 3). More specifically, Bruner says,

> although meanings are "in the mind," they have their origins and their significance in the culture in which they are created...It is this cultural situatedness of meanings that assures their...communicability...[thus] knowing and communicating are in their nature highly interdependent. (p. 3)

Unlike an information-processing, data-driven model, a culturalist view of intelligence is a way of knowing that would call for re-envisioning the nature of intelligence, away from a mode of observer consciousness, and toward a recognition that knowing and being are intersubjectively cons-tructed—a perspective on intelligence more resonant with an ecospiritual praxis for education.

A process view of reality is increasingly applied to the ways that scholars are envisioning curriculum, not meant as a model or a method, but as a way of questioning how curriculum might be re-conceived (Doll, 1993). A process view of curriculum would move from the personal and social to the ecological and cosmological and would require a more holistic view of schooling—valuing and making connections between the body, emotions, mind, and spirit. Such an educational frame would be ecological in its broadest sense, recognizing and evoking a deep respect within children for the larger, organic and cosmic processes so vital to our existence. Such a vision would be a departure from the cause-effect, linear models put forth by behaviorists and social efficiency educators which have traditionally separated and isolated the disciplines. This open-systems, eco-cosmic view of curriculum would allow for a "complex interplay between openness and closure at a number of levels" (p. 58) providing the cohesive-ness of an integrated structure, yet allowing for interactions and arisings which bubble-up from their situatedness within a nexus of relations. It would clear a space for possibilities to come forth, beyond dualistic frameworks of either openness or closure, toward clearings wherein new levels of complexity and dynamic interaction may emerge (Doll, 1993, pp. 58–68).

Quantum mechanics has taught us that all of the things in the universe that appear to exist independently are actually parts of one "all-encompassng organic pattern, and that no parts of that pattern are ever really separate from it or from each other" (Zukav, 1979/1986, p. 48). Moving with the cycles of tides, seasons, cells, and celestial bodies is a more earth-based approach for viewing social systems composed of beings who are biologically constituted for life within systems based on relation-ality and complexity. The new wave of the future is not fragmentary either/or thinking of dualistic, patriarchal worldviews. *Both/and* thinking is replacing either/or: both complexity and simplicity; both chaos and order; both dynamic openness and an organizing gravity; unity and diversity; the one and the many. With Doll (1993), I feel that this approach may assist us in devising "more relational or ecological...ways to view and interact with our environment" (p. 65), and to vitalize communities within schools.

Vitalization of this sort suggests the integrative knowing of body, heart, mind, and spirit, rather than separating the ways we approach education into the "intellectual and cultural binaries that ensnare creative thinking in the contemporary context" (Smith, 1996, p. 6.). David Jardine's (2000) notion of "ecopedagogical reflection" suggests a reawakening of the "sense of intimate connection between ecological awareness and pedagogy" (p. 87). He bases this connection on an understanding that the ongoing and creative nature of life is "always already intimately pedagogic at its heart" and "to the extent that pedagogy...usher[s] children into those understandings of the Earth's ways required for life to go on in a full and healthy and wholesome and sustainable way, it is already intimately ecological at its heart" (p. 48). In watching the world around us, it is clear that a linear way of educating children only limits their capacities to deal fully with life's messiness, life's joy—the abundant possibilities available when living isn't fragmented into preestablished boundaries based only on "clear and solid and fixed foundations" (p. 49). Opening our eyes, it is readily apparent "that there is no such securable ground to the living practices of human life, only on-going, shifting, ambiguous nests or communities of interrelations which are constantly in need of re-newal, re-generation, re-thinking" (p. 50).

There is no separation between the process of creation and the part that humans play within that process. And, we are creators on this planet even while we are also implicated in its destruction. We can choose to create "life" and can also choose to destroy it, remembering that choosing not-to-choose is also a choice. And by our lives and our actions do our creations come to bear. A choice to wake up is always in our midst, just below the surface, present in the now-moment. It is not some grand enterprise of awakening, rather, always in small increments—because this work is never finished, but always made new in the local places. *This* relationship, *this* piece of ground, *this* owl who is watching me. It is ever-now that the work of ecospiritual praxis takes place, locally, within the conscious choice to create: our "strategic intentions." Ecospiritual awareness suggests an integration of our varied capacities to know and experience life. It

> involves deeply spiritual attention required to be mindful of each gesture, each breath, and the cherishing of the interdependencies and inevitabilites that house us…but it requires also an ability to read and become mindful of the violations and compromises of such attention, of the violences and severences out of which so much of our lives and the lives of our children are built…Without these two moments, ecopedagogical reflection becomes a romanticism that refuses to face where it actually is and the threads of culpability that bind it here and here and here. (Jardine, 2000, p. 34)

Without critical reflection into the realities of the world as it exists, a narrowly conceived spirituality can risk the "escapism" of a kind of fly-away-theology, often taking place in a "fragmented or commodified way" (Mies & Shiva, 1993, p. 19). For instance,

> [t]hose interested in a sort of oriental spiritualism rarely know …how people in, for example, India live or even the socio-economic and political contexts from which fragments—such as yoga or tai-chi—have been taken. It is a kind of luxury spirituality. (p. 19)

Spirituality, divorced from the fleshly matters of the material world, is incomprehensible to many Third World women for whom

> the earth is a living being which guarantees their own and their fellow creatures survival. They respect and celebrate Earth's sacredness and resist its transformation into dead, raw material for industrialism and commodity production. It follows therefore that they also respect the diversity and the limits of nature which cannot be violated if they want to survive. (p. 18)

An ecospiritual praxis, then, is not complete without a consistent mode of reflection, along with a willingness to act toward the re-creation of the world around us, always in the local.

Rites of Passage and Responsibility

John P. Miller (2000) maintains that if "we can see ourselves as part of the web, there is less chance that we will tear the web apart" (p. 5). Helping young people grasp their fundamental connection within a larger system of relationships, or their "situatedness,"encourages within them a cultivation of "their ties to others and the forms of obligation, responsibility, and support associated with those relationships" (Smith & Williams, 1999, p. 9). Ecospiritual praxis calls for a renewing of the relationships among the people who inhabit classrooms. In this regard, the teacher/student relation can be an appropriate and useful ground for teaching and learning.

Relationship is always dynamic, it never stays the same. Because it shifts and changes with each negotiation, each interaction between and among participants, we say that it is in process. The teacher/student relationship is one which is always in process and is ideally a two-way negotiation, unlike Freire's (1970) critique of the "banking concept of education" (p. 7). Palmer (1983) reminds us that the root word meaning *"to educate"* comes from the Old English "to draw out" (pp. 81–82). He suggests that "the teacher's task is not to fill the student with facts but to evoke…[what] the student holds within" (p. 43), to bring forward the potential from within each child (Dewey, 1956). It suggests a partnership, a mutual and active relation between the educator and the child, rather than the passive child-as-receiver/teacher-as-transmitter (Kesson, 1994, pp. 2–6) models discussed earlier. Furthermore, the emphasis should move beyond the stress on personal individuality to the importance of learning to balance the individual's place within the larger social and biological world (Bowers, 1991). As relational beings, one of the most vital qualities that a human may have is an ability to respond—to others and to the surrounding environment. The word *respond* comes from the Latin *spondere*, meaning to make a solemn promise, pledge, betroth; from the Greek it means "offering," (*American Heritage Dictionary*, 1996, p. 1537) and it suggests a relation that is deeply felt, an interaction conducive to coming into partnership as a creative force. Teaching children to become aware as response-able partners within relationships can be a beginning road toward helping them see that they have response-ability within a world community much larger than themselves.

The teacher/student relationship is one of the guiding themes that runs through a foundations course I teach within a university department of curriculum and instruction. We begin building that relationship from day one, drawing on principles of participatory education (Jennings & Purves, 1991) to create a classroom *ethos* from which we work for the duration of the course. The time spent in early efforts at creating a good foundation among us begins to frame the relationship we build upon for the entire semester.

We start by examining "what makes a good teacher," brainstorming a long list of qualities based on their prior knowledge. I am also allowed to contribute. The list includes qualities such as: interesting, fast-moving, makes connections, learner-centered, informative, initiates your personal knowledge from within, well-prepared, energetic, compassionate. Then we repeat the process based on qualities they believe make a good student, for example, being: considerate, interested in learning, willing learners, well-disciplined, serious about their work, kind to others, on time, well-prepared, active in class discussions. Once any after-thoughts are added to our two long columns, I ask them if these qualities would "apply to any classroom, or only to elementary classrooms?" They usually agree that they would be good qualities to find in any classroom. "If that is the case," I offer, " what if we tried an experiment? What if we established these as guidelines for *this* course? Would we not have, in effect, begun creating this class together?"

I suggest that we keep the large piece of poster board with our list and post it somewhere. "It could be a guiding frame, an ethos to work toward." I define "ethos" on the board, "The disposition, character, or fundamental values peculiar to a specific person, people, culture, or movement" (*American Heritage Dictionary*, 1996, p. 631). If they acknowledge that the word applies, we title it *Guiding Ethos* and I continue. "I would offer this—that we say that this is not a *closed* document, but that we can add to it when something comes up that we want to address. What do you think about that?" They seem to like the idea.

"One thing comes up for me as I'm thinking about it. This is a community of teachers we're building here, and communities are made up of relationships. One thing I notice that usually makes my relationships better is honesty. When I'm honest with the people in my life, and they're honest with me, things are usually much more healthy in the long run. So, I'd like to add honesty to the list." If they agree, I do that.

"Another thing I notice, though—and it relates to honesty—is consideration. If that element is missing, sometimes honesty can be hurtful or do some amount of harm. So it seems to me that with honesty, you need consideration. Consideration is an important element for making it a safe space to be honest in." If they agree, I add consideration. We talk about what consideration means and then I add, to be considerate is to be "mindful" of the other person. One way to do that is to realize that the "truth" that you're being honest about, is the truth for you—but not necessarily for them—and you recognize and acknowledge that. You have enough respect for them to see that their truth may be different from your truth and that that's okay because no one gave *me* the right to judge what is the truth for *you*. Each one of us does that for ourselves." (This begs a discussion about what "truth" is and if it exists separate from negotiated meaning. I make a note of that for future discussion.)

And so it is with the creation of this document. Through it, we set a conscious intention for the relationship we will engage within for the next three months. It constitutes a ceremony, a *rite* through we may *pass* into a world of our own creation. And we are left with an artifact to keep the process—and the subsequent ethos—alive among us. Also, the practice demonstrates for students how they may establish a personal ethos with their own future students. For very young children, it might be to engage them in creating classroom "rules," a way of giving young ones ownership of the day-to-day mores guiding classroom life. They are basically some ethical principles to help them learn consideration and fairness as they begin to navigate within a social setting for the first time. As well, we have laid a foundation for guiding the rest of the course. It is something we refer to again and again and build upon as we go. Ideal is that they come to view it as their own document as much as mine. They are encouraged to bring it to the group's attention when we are not being consonant with our ethos or when they feel something important needs to be added. This guiding ethos, and what it represents, has the potential to become something larger than any of us individually, a larger frame within which we may become a part. And it opens the ground for honest communication, which always brings risk, so it is a first step toward establishing a space wherein we may interact openly, guided by some simple parameters that help reassure us that we intend to make one another feel safe within the relationship.

Commitment toward Intentionality

An ecospiritual praxis suggests the living of life with deliberateness, as if every moment were meaningful and one could make a difference: commitment comes to mind. Commitment suggests agency—action toward a goal or desired-for dream. Community has the same root and would move the commitment beyond the self into the sharing interaction of the social.

Further, educating young people as to their interdependence with all forms of life is initiating them into a cosmological community of which they are already a part. In Orr's (1994) words:

> We are of the earth...We live in the cycle of birth and death, growth and decay. Our bodies respond daily to rhythms of light and darkness, to the tug of the moon, and to the change of the seasons. (p. 204)

Picking up an etymological thread once again, Orr feels that what should be "drawn out" in the educating process is "our affinity for life" (p. 205). An education that builds on that affinity could bring us to the kind of personal and social "awakenings" to life on which to base "humane and sustainable societies," the kind of "possibilities and potentials that lie largely dormant in the industrial-utilitarian mind" (p. 205).

Engaging the "spirit" of living within the social milieu of human classrooms implies a fostering of awareness within children of their capacities for "wide-awakeness," "imaginative action," and "consciousness of possibility," as well as "multiple conceptions of what it is to be human and alive" (Greene, 1995, p. 43). If there is any hope toward transforming our world from the dualistic, power-over frame of fragmentary thinking, to a more inclusive and relationally framed view, it will lie in the hands of individuals-in-community working at local sites. Rather than an emphasis on *either* individual *or* community (another dualism), *both* individual *and* community connotes a continuum of interactions that constitute a part-to-whole relation. Susie Gablik (1991) has said that

> the source of creativity in society is the person.... Both the problem and the level at which the solution emerges are manifested initially in the individual, who is also an organ of the collective. What happens in the individual is typical of the total situation and is the place where future solutions emerge. (pp. 22–23)

Once again, intentionality comes to mind. What is commitment and how does it tie in with the notion of *choice* drawn along through this discussion? What moves us toward a decision, a centering on, a grounding into some goal, some desired-for dream. How do we summon up the inner resources required of us to act? From where do we draw the strength that constitutes commitment? agency? decision for change? for creation? Is it something that just *happens to us*, willy-nilly, when we become caught up in some frenzy of momentum going on in our environment? Or can the choice be made? Isn't choice somehow a fuel toward the realization of agency? A choice to wake up, a choice to say "'no' to those elements of our lives and our ways that are unsustainable, that befoul our nest" (Jardine, 2000, p. 63), a choice to embrace those "others" who are "different in kind," but who, in their difference, offer possibilities for bridging new relations.

It is the human capacity *to choose* that is one of the potentials ever-present in the now-moment, just beneath the surface. There are those nexus of local places and particular people through which decisions can be made, seizing opportunities for actions, great or small. Actions contribute to change, and the direction that change carries us lies within the conscious choosing, strategic intentions. For it is the inner cultivation of a personal journey for teacher and student—toward the ability to act in ways beneficial to "life" on the planet—that I believe comprises education's greatest potential. I return to Delpit's (1997) suggestion that commitment comes from the recognition that we have a place, indeed, a purpose within something which is larger than ourselves: a relationship, a family, a community, an ecosystem. Further, bell hooks (1994) says that "teachers must be actively committed to a process of self-actualization that promotes their own well-being, if they are to teach in a manner that empowers students" (p. 15). She uses the term "sacred" when she speaks of teaching as going beyond merely sharing information but also sharing in the "intellectual and spiritual growth of our students…in a manner that respects and cares for [their] souls…where learning can most deeply and intimately begin" (p. 13). With Krall (1994), I believe that

> [o]ur greatest challenge will be to replace bureaucratic, institutional, rational, and arbitrary worldviews with those grounded in ecological wisdom and responsibility, where difference is played out in healthy social contexts that are dynamic and pluralistic. (p. 15)

To ever think that the work of ecospiritual praxis can affect children in schools we must begin within ourselves as teachers, as human beings. This work is no formula, or recipe, or method, but is a path, a journey that can only be one's own. By living this journey toward what Smith (1996) calls "being awake to the way that sustains us, we face ourselves" (pp. 9–11) and work for our own understandings of what is important to enrich and fulfill us along the way.

Notes

1. Mies and Shiva (1993) say "globalism" through the lens of "capitalist patriarchy" is the "global reach of capital to embrace all the word's resources and markets" (p. 109). They describe the term 'global' in "global order" as the "global domination of local and particular interests...subsuming the multiple diversities of economies, cultures and of nature under the control of a few multinational corporations (MNCs), and the superpowers that assist them in their global reach through 'free' trade, structural adjustment programs and, increasingly, conflicts, military and otherwise" (p. 9). Subsequently, rather than communities cooperating with one another and with the land, the "homogenization processes of development...[cause] positive pluralities [to] give way to negative dualities, each in competition with every 'other'...characteriz[ing] situations where nothing is sacred but everything has a price" (pp. 111–112).

2. Among things that I would add are sexism, competitive individualism, and anthropocentrism, all of which lead to personal and cultural alienation and also to the devastation of the ecosystem.

Toward a Radical
Re-cognition of Life

So, why do we resist the movement toward "life?" It happens all around us. It is February, the threshold of spring here in the south. There is a rejuvenation all around. I can feel it as I walk back from the mail-box—down the hill along the line of brown twiggy elderberries. There is no green as yet, but in the air is a freshness, an inherent seed of knowing what is to be. One can almost "feel" the movement toward life, toward renewal; recognize a stirring, a circulation begin within the stubby branches—a gathering, a rising, a burgeoning forth. It will come from the ends, at first, a popping out of green. And then, it will spread into a profusion of burgeonings and buddings, into an array of every shade of green imaginable.

This is our life—this is all around, it comes to us regardless, ofttimes unaware. We do not create it nor control it, rather we *are* it and it is us. We are a part of it. As a species, we too are sustained within the cyclical movements which living affords, and we experience, also, this freshening, this renewal. Is there no way to allow this movement toward life, also, into our institutions? How do we build communities of learning which could bring in a sense of the spiritual without systematizing "magic" into mechanization? How do we give voice within schools to an acknowledg-ment of the awe and wonder which mark our connection with some larger sense of order. How do we suggest to young people that life has meaning, has purpose, that there is some context for their lives which has worth and value? How do we show them what it means to *live a commitment*? It will require a constant mode of reflecting and acting to continue the generation of new and more interesting ways of seeing, of thinking, of being, so that we *are* the journey and the journey is us; alive in the moment of it, engaged in the 'present' of it. So that when we are sitting, we *know* we are sitting—when we are working, we *know* we are working—because we are *engaged* within the present moment of the experience. And as teachers, it is in learning to reflect, to become aware, to become engaged to life, that we may mirror these things for those whose lives we affect and whose lives affect our own.

References

Abram, D. (1996). *The spell of the sensuous: Perception and language in a more-than-human world*. New York: Vintage.

Abram, D. and Jardine, D. W. (2001). All knowledge is carnal knowledge: A correspondence. In B. Hocking, J. Haskell, and W. Linds (Eds.), *Unfolding bodymind: Exploring possibility through education* (pp. 315–323). Brandon, VT: Foundation for Educational Renewal.

Adams, C. (1993). *Ecofeminism and the sacred*. Continuum: New York.

American heritage dictionary of the English language (3 ed.). (1996). (A. H. Soukhanov, Ed.). Boston: Houghton Mifflin.

Aoki, T. T. (1993). Legitimating lived curriculum: Towards a curricular landscape of multiplicity. *Journal of Curriculum and Supervision, 8*(3), 255–268.

Bateson, G. (1979). *Mind and nature: A necessary unity*. New York: Dutton.

Belenky, M. F., Clinchy, B. M., Goldberger, N. R., & Tarule J. M. (1986). *Women's ways of knowing: The development of self, voice, and mind*. New York: HarperCollins.

Bergland, B. (1994). Postmodernism and the autobiographical subject: Reconstructing the other. In K. Ashley, L. Gilmore, & G. Peters (Eds.), *Autobiography & postmodernism* (pp. 130–166). Amherst: University of Massachusetts Press.

Bergson, H. (1911). *Creative evolution*. New York: Holt.

Berman, M. (1981). *The reenchantment of the world*. New York: Cornell University Press.

Bernstein, R. (1991). *The new constellation: The ethical-political horizons of modernity/postmodernity*. Cambridge: MIT.

Berry, T. (1988). *The dream of the earth*. San Francisco: Sierra Club.

Bloom, L. A. and Munro, P. (1996). Conflicts of selves: Nonunitary subjectivity in women administrators' life history narratives. In J. A. Hatch and R. Wisniewski (Eds.), *Life history and narrative* (pp. 99–112). London: Falmer.

Bohm, D. (1983). *Wholeness and the implicate order*. London: Ark.

————. (1985). *Unfolding meaning: A weekend of dialogue with David Bohm.* London: Ark.

Bookchin, M., (1982). *The ecology of freedom: The emergence and dissolution of hierarchy.* Palo Alto: Cheshire.

————. (1991). *Defending the earth: A dialogue between Murray Bookchin and Dave Foreman.* Boston: South End Press.

Bowers, C. A. (1987). *Elements of a post-liberal theory of education.* New York: Teachers College Press.

————. (1991). An open letter to Maxine Greene on "The problem of freedom in an era of ecological interdependence." *Educational Theory, 41,* 325–330.

————. (1993). *Critical essays on education, modernity, and the recovery of the ecological imperative.* New York: Teachers College.

————. (1995). *Educating for an ecologically sustainable culture:Rethinking moral education, creativity, intelligence, and other modern orthodoxies.* Albany: State University of New York Press.

————. (1997). *The culture of denial: Why the environmental movement needs a strategy for reforming universities and public schools.* Albany: State University of New York Press.

————. (2001). *Educating for eco-justice and community.* Athens: University of Georgia Press.

Bowers, C. A.& Flinders, D. G. (1990). *Responsive teaching.* New York: Teachers College.

Bragg, E. (1999). Deep ecology. [On-line]. (excerpts from *Towards an ecological self: Individual and shared understandings of the relationship between self and the natural environment.* Unpublished doctoral dissertation, 1995, James Cook University of North Queensland, Townsville). Available: <http://forests.org/ric/seed/deep-eco>

Brown, B. (1993). Toward a Buddhist ecological cosmology. In M. E. Tucker and J. A. Grim (Eds.), *Worldviews and ecology.* (pp.124–137). Lewisburg: Bucknell.

Brill, S. B. (1995). *Wittgenstein and critical theory: Beyond postmodern criticism and toward descriptive investigations.* Athens: Ohio University Press.

Bruner, J. (1996). *The culture of education*. Cambridge: Harvard University Press.

Burbules, N. (1991). Two perspectives on reason as an educational aim: The virtues of reasonableness. *Proceedings of the 47th Annual Meeting of the Philosophy of Education Society*, 215–224.

Butterfield, S. (1994). *Black autobiography in America*. Amherst: University of Massachusetts.

Capra, F. (1975/1991). *The Tao of physics*. Boston: Shambala.

———. (1996). *A new scientific understanding of living systems: The web of life*. New York: Doubleday.

Chadwick, D. (1999). *The life and Zen teaching of Shunryu Suzuki*. New York: Random House.

Chief Seattle (1854). In S. Jeffers, (Ed.), *Brother eagle, sister sky: A message from Chief Seattle*. New York: Penguin.

Cheney, J. (1999). The journey home. In A. Weston (Ed.), *An invitation to environmental philosophy* (pp. 141–168). New York: Oxford University Press.

Collard, A. (with Contrucci, J.). (1989). *Rape of the wild: Man's violence against animals and the earth*. Bloomington: Indiana University Press.

Daly, M. (1978). *Gyn/Ecology: The metaethics of radical feminism*. Boston: Beacon Press.

Davies, P. (1992). *The mind of God: The scientific basis for a rational world*. New York: Simon & Schuster.

Deleuze, G. (1977/1987). *Dialogues*. London: Athlone Press.

Delpit, L. (1997). Public lecture. Louisiana State University. April, 7, 1997.

Derrida, J. (1993). *Aporias*. (T. Dutoit, Trans.). Stanford: University Press.

Devall, B. (1988). *Simple in means, rich in ends: Practicing deep ecology*. Salt Lake City: Gibbs-Smith.

Dewey, J. (1897). My pedagogic creed. In Encyclopedia Britannica (Eds.), *Annals of America: Vol. 12. Populism, imperialism, and reform* (pp. 125–130). Chicago: Encyclopedia Britannica.

———. (1902). *The child and the curriculum*. Chicago: Chicago University Press.

————. (1934). *Art as experience.* Paragon: New York.

————. (1956). *The school and society and The child and the curriculum.* Chicago: Chicago University Press.

Dingler, J. (1999). Ph.D. research. [On-line]. Available: <http://userpage.iu- bernn.de/~jdingler/ecofem.num>

Doll, M. A. (1995). *To the lighthouse and back: Writings on teaching and living.* New York: Peter Lang.

Doll, W. (1993). *A post-modern perspective on curriculum.* New York: Teachers College.

————. (1997). *The Spirit of Education.* Paper presented at Contemporary Curriculum Dialogues, University of Calgary, Alberta, Canada.

————. (2002). Struggles with spirituality. In D. Carlson & T. Oldenski (Eds.), *Educational yearning: The journey of the spirit and democratic education.* New York: Peter Lang.

Dreyfus, H. L., & Rabinow, P. (1982). *Michel Foucault: Beyond structuralism and hermeneutics.* Chicago: Chicago University Press.

Eckersley, R. (1992). *Environmentalism and political theory: Toward an ecocentric approach.* New York: University Press.

Eisler, R. (1987). *The chalice and the blade.* San Francisco: Harper & Row.

————. (1990). The Gaia tradition and the partnership future: An ecofeminist manifesto. In I. Diamond & G. Feman Orenstein (Eds.), *Reweaving the world: The emergence of ecofeminism* (pp. 15–22). San Francisco: Sierra Club.

Emerson, R. W. (1965). The journal, V, 411–412; In Fremantle (Ed.), *The Protestant mystics* (pps. 189–190), Boston: Mentor.

Foucault, M. (1973). *The order of things:An archaeology of the human sciences.* New York: Random House.

————. (1978). *The history of sexuality.* New York: Random House.

————. (1979). *Discipline & punish: The birth of the prison.* New York: Random House.

Fox, M. (1983). *Original blessing.* Santa Fe: Bear & Co.

————. (1988). *The coming of the cosmic christ: The healing of Mother*

Earth and the birth of a global renaissance. San Francisco: Harper & Row.

————. (1995). *Wrestling with the prophets: Essays on creation spirituality and everyday life*. San Francisco: HarperCollins.

Freire, P. (1970/1993). *Pedagogy of the oppressed*. New York: Continuum.

Fromm, E. (1955). *The sane society*. New York: Holt.

————. (1986). *For the love of life*. New York: Macmillan.

Furlong, J. (2000). Intuition and the crisis in teacher professionalism. In T. Atkinson & G. Claxton (Eds.), *The intuitive practitioner: On the value of not always knowing what one is doing* (pp. 16–31). Philadelphia: Open University Press.

Gablik, S. (1991). *The reenchantment of art*. New York: Thames and Hudson.

Gardner, H. (1993). *Frames of mind: The theory of multiple intelligences*. New York: Basic Books.

Garrison, J. (1998). *Dewey, Coleridge, and Education for Spirituality*. Paper presented at the annual meeting of the American Educational Research Association, San Diego, CA.

Gergen, K. J. (1991). *The saturated self: Dilemmas of identity in contemporary life*. United States: HarperCollins.

Gilmore, L. (1994). *Autobiographics: A feminist theory of women's self-representation*. Ithaca: Cornell University.

Goodman, K. (1986). What's whole in whole language. Portsmouth, NH: Heinemann.

Goodrich, N. L. (1989). *Priestesses*. New York: Franklin Watts.

Greene, M. (1995). *Releasing the imagination*. San Francisco: Jossey-Bass.

————. (1996). Public lecture. Louisiana State University. June 27, 1996.

Griffin, S. (1990). Curves Along the Road. In I. Diamond & G. Feman Orenstein (Eds.), *Reweaving the world: The emergence of ecofeminism* (pp. 87–99). San Francisco: Sierra Club.

Habermas, J. (1976/1979). *Communication and the evolution of society*. (T. McCarthy, Trans.). Boston: Beacon Press.

Hanh, T. N. (1992). *Touching peace: Practicing the art of mindful living.* Berkley: Parallax.

———. (1996). *Be still and know: Reflections from living Buddha, living Christ.* New York: Riverhead.

Haraway, D. (1991). *Simians, cyborgs, and women: The reinvention of nature.* New York: Routledge.

Harding, S. (1991). *Whose science? Whose knowledge? Thinking from women's lives.* Ithaca: Cornell.

Harvey, D. (1989). *The condition of postmodernity.* Oxford: Basil Blackwell.

Harvey, E. D. & Okruhlik, K. (1992). *Women and Reason.* Ann Arbor: University of Michigan Press.

Hayles, K. N. (1991). Chaos and order: Complex dynamics in literature and science. Chicago: Chicago University Press.

Hekman, S. J. (1990). *Gender and knowledge: Elements of a postmodern feminism.* Boston: Northeastern University Press.

Hocking, B., Haskell, J. and Linds, W. (2001). Re-imag(e)ining Worlds through education: An overview of the book and its influences. In B. Hocking, J. Haskell, and W. Linds (Eds.), *Unfolding bodymind: Exploring possibility through education* (pp. xvi–xxxvii). Brandon, VT: Foundation for Educational Renewal.

Hogan, L. (1996). The kill hole. In R. S. Gottlieb, (Ed.), *Sacred earth:Religion, nature, environment* (pp. 37–40). New York: Routledge.

hooks, b. (1994). *Teaching to transgress.* New York: Routledge.

Horkheimer, M. & Adorno, T. (1972). *Dialectic of enlightenment.* New York: Herder.

Hoy, D. (1988). *After Foucault: Humanistic knowledge, postmodern challenges.* New Brunswick: Rutgers.

Huebner, D. (1995). Education and Spirituality. *JCT: An interdisciplinary journal of curriculum studies,11*(2), 13–34.

———. (1999). *The lure of the transcendent.* Mahwah, NJ: Erlbaum.

Jacobs, D. T. (1998). *Primal awareness: A true story of survival, transformation, and awakening with the Raramuri shaman of Mexico.* Rochester, VT: Inner Traditions.

Jacobs, D. T. & Jacobs-Spencer, J. (2001). *Teaching virtues: Building character across the curriculum.* London: Scarecrow Press.

Jaggar, A. M. (1989). Love and knowledge: Emotion in feminist episte-mology. In A. Jaggar & S. R. Bordo, (Eds.), *Gender/body/knowledge: Feminist reconstructions of being and knowing* (pp. 145–171). New Brunswick, New Jersey: Rutgers University Press.

James, W. (1907/1995). *Pragmatism.* New York: Dover.

Jardine, D. W. (2000). *"Under the tough old stars": Ecopedagogical es-says.* Brandon, VT: Foundation for Educational Renewal.

Jay, M. (1973). *The dialectical imagination: A history of the Frankfurt school and the institute of social research, 1923–1950.* Boston: Little, Brown & Co.

Jencks, C. (Ed.), (1992). *The post-modern reader.* New York: St. Martin's Press.

Jennings, E. M. & Purves, A. C. (Eds.), (1991). *Literate systems and in-dividual lives: Perspectives on literacy and schooling.* New York: University Press.

Kaza, S. (1993). Acting with compassion: Buddhism, feminism, and the environmental crisis. In C. J. Adams (Ed.), *Ecofeminism and the sacred* (pp. 50–69). New York: Continuum.

Keller, C. (1993). Talk about the weather: The greening of eschatology. In C. J. Adams (Ed.), *Ecofeminism and the sacred* (pp. 30–49). New York: Continuum.

Keller, E. (1985). *Reflections on gender and science.* New Haven: Yale University Press.

Kesson, K. (1994). Recollections: An introduction to the spiritual dimen-sions of curriculum. *Holistic Education Review, 7* (3), 2–6.

King, Y. (1990). Healing the wounds: Feminism, ecology, and the nature/culture dualism. In I. Diamond & G. F. Orenstein (Eds.), *Reweaving the world: The emergence of ecofeminism* (pp. 106–121). San Francisco: Sierra Club.

Kohli W. (1984a). Reflections of a Critical Educator. *Kairos, 1*(3), 32–37.

————. (1984b). *Toward hermeneutic competence: The empowerment of teachers.* Unpublished doctoral dissertation, Syracuse University, New York.

————. (1991a). Humanizing education in the Soviet Union: A plea for caution in these postmodern times. *Studies in Philosophy and Education, 11*(1), 51–63.

————. (1991b). Postmodernism, critical theory and the "new" pedagogies: What's at stake in the discourse? *Education and Society, 9*(1), 39–46.

————. (1991c). Critical hermeneutics: Towards a philosophical foundation for the empowerment of teachers. In *Philosophy of Education Society 1991,* (pp. 119–129). Urbana: Philosophy of Education Society.

————. (1993). Raymond Williams, affective ideology, and counter-hegemonic practices. In M. W. Apple (Ed.), *Views beyond the border country* (pp. 115–132). New York: Routledge.

————. (1995). Educating for emancipatory rationality. In W. Kohli (Ed.), *Critical conversations in philosophy of education* (pp. 103–115). New York: Routledge.

————. (1996). Teaching in the danger zone. *The International Journal of Social Education, 11*(1), 1–16.

Krall, F. R. (1994). *Ecotone: Wayfaring on the margins.* New York: University Press.

LaDuke, W. (1996). Indigenous Mind. *Resurgence* [On-line]. Available: <http://www.gn.apc.org/resurgence/articles/laduke.htm>

Laszlo, E. (1972). *The systems view of the world.* New York: Braziller.

Lee, D. (1959). *Freedom and culture.* Englewood Cliffs, NJ: Prentice-Hall.

Leiss, W. (1972). *The domination of nature.* New York: Braziller.

Leopold, A. (1949). *A Sand County almanac and sketches here and there.* London: Oxford University Press.

Longino, H. E. (1989). Can there be a feminist science? In A. Garry & M. Pearsall (Eds.), *Women, knowledge, and reality: Explorations in feminist philosophy* (pp. 203–216). Boston: Unwin Hyman.

References

Lydon, A. (1995). An ecozoic cosmology of curriculum and spirituality. *Journal of curriculum theorizing: An interdisciplinary journal of curriculum studies. 11*(2), 67–86.

Lyotard, J. (1979). *The postmodern condition: A report on knowledge.* Minneapolis: University of Minnesota Press.

Macdonald, J. B. (1964/1995). *Theory as a prayerful act: The collected essays of James B. Macdonald.* New York: Peter Lang.

Macy, J. (1991a). *World as lover, world as self.* Berkley: Parallax Press.

———. (1991b). *Mutual Causality in Buddhism and general systems theory.* Albany: New York State.

Margulis, L. & Lovelock, J. E. (1989). Gaia and geognosy. In M. Rambler, L. Margulis, & R. Fester (Eds.), *Global ecology* (pp. 1–30). San Diego: Academic Press.

McDonough, W. (2001, February). *Keynote address.* Paper presented at the meeting of the Foundation for the Carolinas, Charlotte, N.C.

Merchant, C. (1980). *The death of nature: Women, ecology and the scientific revolution.* San Francisco: HarperCollins.

———. (1996). *Earthcare: Women and the environment.* New York: Routledge.

Mies, M. & Shiva, V. (1993). *Ecofeminism.* Spinifex: Victoria, Australia.

Miller, J. (1997, March). *Un-telling teachers' stories as curriculum history.* Paper presented at the American Educational Research Association Conference, New York, NY.

Miller, J. P. (2000). *Education and the soul: Toward a spiritual curriculum.* Albany: SUNY Press.

Morrison, G. S. (1997/2000). *Teaching in America.* Boston: Allyn & Bacon.

Munro, P. (1996, April). *Catching the "true" history: Poststructuralism, gender and curriculum history.* Paper presented at the American Educational Research Association Conference, New York, NY.

———. (1998). *Subject to fiction: Women teachers' life history narratives and the cultural politicas of resistance.* Buckingham: Open University Press.

Noddings, N. (1992). *The challenge to care in schools: An alternative approach to education.* Teachers College: Columbia.

Olney, J. (Ed.), (1980). Autobiography: Essays theoretical and critical. Princeton, NJ: Princeton University Press.

O'Neill (1976). *On critical theory.* New York: Seabury.

Orr, D. W. (1992). *Ecological literacy: Education and the transition to a postmodern world.* New York: State University of New York Press.

———. (1994). *Earth in mind: On education, environment, and the human prospect.* Washington: Island Press.

Pagano, J. (1990).*Exiles and communities: Teaching in the patriarchal wilderness.* Albany, NY: State University of New York Press.

Palmer, P. (1983). *To know as we are known: A spirituality of education.* San Francisco: Harper & Row.

———. (1998). *The courage to teach: Exploring the inner landscape of a teacher's life.* San Francisco: Jossey-Bass.

Payne, D. G. (1996). *Voices in the wilderness: American nature writing and environmental politics.* Hanover, NH and London: University Press of New England.

Pinar, W. (1974). *Heightened consciousness, cultural revolution and curriculum theory: The proceedings of the Rochester conference.* Berkeley: McCutchan.

———. (1975). *Curriculum theorizing: The reconceptualists.* Berkeley: McCutchan.

———. (1985). *Autobiography, politics and sexualtiy: Essays in curriculum theory 1972–1992.* New York: Peter Lang.

———. (1996). *Regimes of reason and male narrative voice.* Unpublished manuscript.

———. (1998). *Queer Theory in Education.* Mahwah, NJ: Lawrence Erlbaum.

———. (1999). After Christianity [Review of the book *Holy Sparks*]. *Educational Researcher, 28*(3), 39–42.

Pinar, W. F. & Grumet, M. R. (1976). *Toward a poor curriculum.* Dubuque: Kendall/Hunt.

Pinar, W. F., Reynolds, W. M., Slattery, P., & Taubman, P. M. (1995). *Understanding curriculum: An introduction to the study of historical and contemporary curriculum discourses*. New York: Peter Lang.

Plumwood, V. (1996). Nature, self, and gender: Feminism, environmental philosophy, and the critique of rationalism. In Karen J. Warren (Ed.), *Ecological feminist philosophies* (pp. 157–180). Bloomington: University Press.

———. (1997). Androcentrism and anthroponcentrism: Parallels and politics. In Karen J. Warren (Ed.), *Ecofeminism: Women, culture, nature* (pp. 327–355). Bloomington: Indiana University Press.

Prigogene, I. & Stengers, I. (1984). *Order out of chaos: Man's new dialogue with nature*. Toronto: Bantum Books.

Quinby, L. (1990). Ecofeminism and the politics of resistance. In I. Diamond & G. F. Orenstein (Eds.), *Reweaving the world: The emergence of ecofeminism* (pp. 122–127). San Francisco: Sierra Club.

Quinn, D. (1992). *Ishmael*. New York: Bantam.

Quinn, M. (2001).*Going out not knowing whither: Education, the upward journey, and the faith of reason*. New York: Peter Lang.

Ramazanoglu, C. (1993). *Up against Foucault: Explorations of some tensions between Foucault and feminism*. New York: Routledge.

Rasberry, G. W. (2001). *Writing research/researching writing: Through a poet's i*. New York: Peter Lang.

Rockefeller, S. C. (1989). John Dewey, spiritual democracy, and the human future. *Cross Currents, 3*, 301–321.

Ruether, R. R. (1996). Ecofeminism: symbolic and social connections of the oppression of women and the domination of nature. In R. Gottlieb (Ed.), *This sacred earth: Religion, nature, environment* (pp. 322–333). New York: Routledge.

Salleh, A. (1997). *Ecofeminism as politics: Nature, Marx and the postmodern*. London: Zed.

Sawicki, J. (1991). *Disciplining Foucault: Feminism, power, and the body*. New York: Routledge.

Siegel, H. (1987). *Relativism refuted: A critique of contemporary epistemological relativism*. Dordrecht, Holland: Reidel.

————. (1991). Two perspectives on reason as an educational aim: The rationality of reasonableness. *Proceedings of the 47th Annual Meeting of the Philosophy of Education Society*, 225–233.

Sequel, M. L. (1966). *The curriculum field: Its formative years*. New York: Teachers College.

Shiva, V. (2000). *Tomorrow's biodiversity*. New York: Thames & Hudson

Sky, M. (1993). *Sexual peace: Beyond the dominator virus*. Santa Fe: Bear & Co.

Smith, D. (1996). Identity, self, and other in the conduct of pedagogical action: An east/west inquiry. *Journal of curriculum theorizing: An interdisciplinary journal of curriculum studies, 12*(3), 6–12.

Smith, G. A. & Williams, D. R. (1999). *Ecological education in action: On weaving education, culture, and the environment*. New York: University Press.

Spretnak, C. (1990). Ecofeminism: Our roots and our flowering. In I. Diamond & G. F. Orenstein (Eds.), *Reweaving the world: The emergence of ecofeminism* (pp. 1–14). San Francisco: Sierra Club.

————. (1991). *States of grace/ The recovery of meaning in the postmodern age: Reclaiming the core teachings and practices of the great wisdom traditions for the well-being of the earth community*. San Francisco: HarperCollins.

————. (1993). Critical and constructive contributions of ecofeminism. In M. E. Tucker and J. A. Grim (Eds.), *Worldviews and ecology* (pp. 181–189). Lewisburg: Bucknell.

————. (1997). *Resurgence of the real: Body, nature, and place in a hypermodern world.* New York: Addison-Wesley.

Starhawk. (1994). Consciousness, politics, and magic. In C. Spretnak, (Ed.), *The politics of women's spirituality: Essays by founding mothers of the women's movement* (pp. 172–184). New York: Doubleday.

Sturrock, J. (1986). *Structuralism.* London: Paladin.

Suzuki, S. (1970/1998). *Zen mind, beginner's mind.* New York: Weatherhill.

Talbot, M. (1991). *The holographic universe*. New York: HarperCollins.

Tyack, D. (1974). *The one best system: A history of American urban education*. Cambridge: Harvard University Press.

Tyler, R. (1949). *Basic principles of curriculum and instruction*. Chicago: Chicago University Press.

Varela, F. J., Thompson, E., & Rosch, E. (1991). *The embodied mind*. Cambridge, Massachusetts: MIT.

Walkerdine, V. (1990). *Schoolgirl fictions*. London: Verso.

Warren, K. (1993). *Ecofeminism and the sacred*. New York: Continuum.

――――. (1996). The power and the promise of ecological feminism. In K. J. Warren, (Ed.). *Ecological feminist philosophies* (pp. 19–41). Bloomington: University Press.

Watts, A. (1957/1989). *The way of Zen*. New York: Vintage.

Weaver, J. (1994). Images and models—in process. In C. Spretnak, (Ed.), *The politics of women's spirituality: Essays by founding mothers of the women's movement* (pp. 250–256). New York: Doubleday.

Weber, M. (1968). *Economy and society: An outline of interpretive sociology*. New York: Bedminster.

――――. M. (1983). The uniqueness of western civilization. In S. Andreski (Ed.), *Max Weber on capitalism, bureaucracy and religion: A selection of texts*. London: George Allen & Unwin. (Original work published 1920 from introduction to *Religionssoziologie*.)

Weston, A. (1999). Epilogue: Going on. In A. Weston (Ed.), *An invitation to environmental philosophy* (pp. 169–196). New York: Oxford University Press.

Wexler, P. (1996). *Holy sparks*. New York: St. Martin's Press.

Willis, G., Schubert, W., Bullough, R., Jr., Kridel, C. & Holton, J. (1993). *The American curriculum: A documentary history*. Westport: Greenwood.

Whitehead, A. N. (1978). *Process and reality: An essay in cosmology*. New York: Macmillan.

Zirbes, L. (1934). Social studies in a new school. *Progressive Education, XI* (1–2).

Zukav, G. (1979/1986). *The dancing wu li masters: An overview of the new physics*. New York: Bantam.

—————. (1989). *The seat of the soul*. New York: Simon & Schuster.

Index

Studies in the Postmodern Theory of Education

General Editors
Joe L. Kincheloe & Shirley R. Steinberg

Counterpoints publishes the most compelling and imaginative books being written in education today. Grounded on the theoretical advances in criticalism, feminism, and postmodernism in the last two decades of the twentieth century, Counterpoints engages the meaning of these innovations in various forms of educational expression. Committed to the proposition that theoretical literature should be accessible to a variety of audiences, the series insists that its authors avoid esoteric and jargonistic languages that transform educational scholarship into an elite discourse for the initiated. Scholarly work matters only to the degree it affects consciousness and practice at multiple sites. Counterpoints' editorial policy is based on these principles and the ability of scholars to break new ground, to open new conversations, to go where educators have never gone before.

For additional information about this series or for the submission of manuscripts, please contact:

Joe L. Kincheloe & Shirley R. Steinberg
c/o Peter Lang Publishing, Inc.
275 Seventh Avenue, 28th floor
New York, New York 10001

To order other books in this series, please contact our Customer Service Department:

(800) 770-LANG (within the U.S.)
(212) 647-7706 (outside the U.S.)
(212) 647-7707 FAX

Or browse online by series:

www.peterlangusa.com